Dennis Hart Mahan

Descriptive Geometry

As Applied to the Drawing of Fortification and Stereotomy ...

Dennis Hart Mahan

Descriptive Geometry
As Applied to the Drawing of Fortification and Stereotomy ...

ISBN/EAN: 9783337269050

Printed in Europe, USA, Canada, Australia, Japan

Cover: Foto ©berggeist007 / pixelio.de

More available books at **www.hansebooks.com**

DESCRIPTIVE GEOMETRY,

AS APPLIED TO

THE DRAWING OF

FORTIFICATION AND STEREOTOMY.

FOR THE USE OF

THE CADETS OF THE U. S. MILITARY ACADEMY.

BY

D. H. MAHAN, LL.D.,

PROFESSOR OF FORTIFICATION, CIVIL ENGINEERING, &C., UNITED STATES MILITARY ACADEMY

NEW EDITION, WITH ADDITIONS.

NEW YORK:
JOHN WILEY & SONS,
15 ASTOR PLACE.
1883.

Trow's
Printing and Bookbinding Co.,
PRINTERS AND STEREOTYPERS,
205-213 *East 12th St.*,
NEW YORK.

PREFACE.

THE subjects of the following pages have been taught orally at the Military Academy for many years; but, for the saving of time, and the convenience of the pupils, it has been thought best to clothe them in a printed dress; and as, in this form, the volume might be found useful in other schools, as an application of descriptive geometry to practical questions, it was also thought well to have it published.

ONE PLANE DESCRIPTIVE GEOMETRY

AS APPLIED TO

FORTIFICATION DRAWING.

1. The method now in general use, among military engineers, for delineating the plans of permanent fortifications, is similar to the one which had been previously employed for representing the natural surface of ground in topographical and hydrographical maps; and which consists in projecting, on a horizontal plane at any assumed level, the bounding lines of the surfaces and also the horizontal lines cut from them by equidistant horizontal planes, the distances of these lines from the assumed plane being expressed *numerically* in terms of some linear measure, as a yard, a foot, &c.

2. **Plane of Reference or Comparison.** The assumed horizontal plane upon which the lines are projected is termed the *plane of comparison* or *plane of reference*, as it is the one to which the distances of all the lines from it are referred, and as it serves to compare these distances with each other and also to determine the relative positions of the lines.

3. **References.** The numbers which express the distances of points and lines from the plane of comparison are termed references. The unit in which these distances are expressed is usually the linear foot and its decimal divisions.

As the position assumed for the plane of comparison is arbitrary, it may be taken either above or below every point of the surfaces to be projected. In the French military service it is usually taken above, in our own below the surfaces. The latter seems the more natural and is also more convenient, as vertical distances are more habitually estimated from below upwards than in the contrary direction. Each of these methods has the advantage of requiring but one kind of symbol to be used, viz: the numerals expressing the ref erences; whereas, if the plane of comparison were so taken that some of the points or lines projected should lie on one side of it and some on the other, it would be then necessary to use, in connection with the references, the algebraic symbols *plus* or *minus* to designate the points above the plane from those below it.

As the distances of all points are estimated from the plane of comparison, the reference of any point or line of this plane will therefore be zero, (0.0); that of any point above it is usually expressed in feet; decimal parts of a foot being used whenever the reference is not an entire number. When the reference is a whole number it is written with one decimal place, thus (25.0); and when a broken number with at least two decimal places, thus (3.70), (15.63). In writing the reference the mark used to designate the linear unit is omitted, in order that the numbers expressing references may not be mistaken for those which may be put upon the drawing to express the horizontal distances between points.

The references of horizontal lines are written along and upon the projections of these lines. All other references are written as nearly as practicable parallel to the bottom border of the drawing, for the convenience of reading them without having to shift the position of the sheet on which the drawing is made.

This method of representing the projections of objects on one plane alone has given rise to a very useful modification of the one of orthogonal projections on two planes, and has been denominated *one plane descriptive geometry;* the plane of comparison being the sole plane of projection; and the references taking the place of the usual projections on a vertical plane. By this modification the number of lines to be drawn is less; the graphical constructions simplified; and the relations of the parts is more readily seized upon, as the eye is confined to the examination of one set of projections alone.

But the chief advantage of it consists in its application to the delineation of objects, like works of permanent fortification, where, from the great disparity of the horizontal extent covered and the vertical dimensions of the parts, a drawing, made to a scale which would give the horizontal distances with accuracy, could not in most cases render the vertical dimensions with any approach to the same degree of accuracy; or, if made to a scale which would admit of the vertical dimensions being accurately determined, would require an area of drawing surface, to render the horizontal dimensions to the same scale, which would exceed the convenient limits of practice. Taking for example an ordinary scale used for drawing the plans of permanent fortifications of *one inch* to *fifty feet*, or the scale $\frac{1}{600}$, the details of all the bounding surfaces can be determined with accuracy to within the fractional part of a foot, whereas a vertical projection to the same scale would be altogether too small for the same purposes.

4. **Point and Right Line.** To designate the position of a point, *Pl.* 1, *Fig.* 1, the projection of the point and its reference are enclosed within a bracket, thus (28.50). This expresses that the vertical distance of the point from the plane of reference is 28 feet and fifty-hundredths of a foot. The position of a right line oblique to the plane of reference is designated by the projection of the line, and the references of any two of its points. Thus in *Fig.* 1 the points *a* and *b*, upon the projection of the right line, with their respective references (25.15) and (28.50), determine the position of the line with respect to the plane of reference.

When the line is horizontal, or parallel to the plane of reference, its projection, with the reference of one of its points, will be sufficient to designate it, and fix its position with respect to the plane of reference. Thus in *Fig.* 1 the reference (25.15), written upon the projection of the line, expresses that the line is horizontal, and 25.15 feet from the plane of reference.

5. For the convenience of numerical calculation, the position of a line, with respect to the plane of reference, is often expressed in terms of the natural tangent of the angle it makes with this plane; but as this angle is the same as that between the line and its projection, its natural tangent can be expressed by the difference of level between any two points of the line, divided by the horizontal distance between the points. Now, as the difference of level between any two points of the line is the same as the difference of the references of the points, and the horizontal distance between them is the same as the horizontal projection of the portion of the line between the same points, it follows, that the natural tangent of the angle which the line makes with the plane of reference is found *by dividing the difference of the references of the points by the distance in horizontal projeciton between them.*

The vulgar fraction which expresses this tangent is termed the *inclination*, or *declivity* of the line. Thus the fraction $\frac{1}{6}$ would express that the horizontal distance between any two points is six times the vertical distance, or difference of their references; the fraction $\frac{2}{3}$, that the vertical distance between any two points is two-thirds the horizontal distance; *the denominator of the fraction*, in all cases, *representing the number of parts in horizontal projection, and the numerator the corresponding number of parts in vertical distance.*

When the position of a line is designated in this way, it is said to be a line whose inclination or declivity is one-sixth, two-thirds, ten on one, &c., or simply, *a line of one-sixth*, &c.

6. Having the declivity of a line, the difference of reference of any two of its points, the projections of which are given, will be found by multiplying the horizontal distance between them by the fraction which expresses this declivity; in like manner the horizontal distance of any two points will be obtained by dividing the difference of their references by this fraction.

To obtain therefore the reference of a point of a line, having its projection, the horizontal distance between it and that of some other known point of the line must be determined from the scale of the drawing by which the horizontal distances are measured; this distance expressed in numbers, being multiplied by the fraction which expresses the declivity of the line, will give the difference of reference of the two points; the required reference of the point will be found by subtracting this product from the reference of the known point, if it is higher than the one sought, or adding if it is lower. Thus let (25.15) be the reference of a known point higher than the one sought; the horizontal distance between the points being 35.75 feet, and the inclination of the line $\frac{1}{10}$; then $35.75 \times \frac{1}{10} = 3.575$ will be the difference of reference of the points, and $25.15 - 3.575 = 21.575$, the required reference. The converse of this shows that the horizontal distance between two points on this line whose difference of reference is 3.575 will be $3.575 \div \frac{1}{10} = 35.75$ feet.

7. When the projection of a line is divided into equal parts, each of which corresponds to a unit in vertical distance, and the references of the points of division are written, it is termed *the scale of declivity of the line.* In constructing the scale of declivity of a line, the entire references are alone put down; one of the divisions of the equal parts being subdivided into tenths, or hundredths if necessary, so as to give the fractional parts of the references corresponding to any fractional part of an entire division.

8. The true length of any portion of an oblique line between two given points is evidently the hypothenuse of a right angle triangle of which the other two sides are the difference of reference of the points, and their horizontal distance.

9. Plane. The position of a plane oblique to the plane of reference may be determined either by the projections and references of three of its points; by the projections and declivity of two lines in it oblique to the plane of reference; or by the projection of two or more horizontal lines of the plane with their references.

The more usual method of representing a plane is by the projections on the plane of reference of the horizontal lines determined by intersecting it by equidistant horizontal planes. These projections are termed *horizontals of the plane*, those usually being taken the references of which are entire numbers.

10. If in a given plane a line be drawn perpendicular to any horizontal line in it, the projection of this line on the plane of reference will be also perpendicular to the projections of the horizontals. The angle of this line with the plane of reference is evidently the same as that of the given plane with it, and is greater than the angle between any other line drawn in the plane and the plane of reference. This line is, on this account, termed *the line of greatest declivity* of the plane.

11. If the scale of declivity of the line of greatest declivity be constructed, it will alone serve to fix the position of the plane to which it belongs, and to determine the reference of any point of the plane of which the projection is given. For since the horizontals are perpendicular to the scale of declivity, the point where the horizontal drawn through the given projection of a point in the plane cuts this line will determine upon the scale the reference of the horizontal, and therefore that of the point.

12. The inclination or declivity of a plane with the plane of reference may be expressed in the same way as the inclination of its line of greatest declivity. Thus *a plane of one-fourth; a plane of twenty on one; a plane of two-thirds*, express that the natural tangents of the angle between the planes and the plane of reference are respectively represented by the fractions $\frac{1}{4}$, $\frac{2}{1}$, and $\frac{2}{3}$.

13. The horizontal distance between any two horizontal lines in a plane, the angle of which is given, can be found in the same way as the horizontal distance between two points of a line, the inclination of which is given, *Art. 7*, by dividing the difference of the reference of the two horizontal lines by the fraction representing the declivity of the plane; in like manner the difference of references of any two horizontal lines will be obtained by multiplying their horizontal distance by the same fraction.

14. To distinguish the scale of declivity, *Pl. 1, Fig. 2,*

from any other line of a plane, it is always represented by two fine parallel lines, drawn near each other, and crossed at the points of division, where the references are written, by short lines which are portions of the corresponding horizontals.

With the foregoing elements the usual problems of the right line and plane can be readily solved.

15. **Problems of the Right Line and Plane.**

Prob. 1, *Pl.* 1, *Fig.* 3. *Having the projections and refer ences of two lines that intersect, to find the angle between them.*

Let *ab* be the projection of one of the lines, the references of two of its points (10.30) and (4.90) being given *cd* the projection of the other line, (10.30), and (5.0) being the references of two of its points; (10.30) being the point of intersection of the two lines.

Find on each of the lines, *Art.* 7, a point having the same reference (7.0). The line joining these two points will be horizontal, and projected into its true length; taking this line as the base of a triangle of which the other two sides are respectively the true lengths of the portions of the two given lines projected between (10.30) and (7.0), *Art.* 7, the angle at the vertex will be the one required.

16. *Prob.* 2, *Fig.* 4. *Through a point to draw a line parallel to a given line.*

Let *c* (7.50) be the projection of the point; *ab* that of the given line of which the two points (7.0) and (9.0) are known.

Through *c* drawing *cd* parallel to *ab*, this will be the projection of the required line; and as its declivity is the same as that of the given line, it will be only necessary to set off from *c* towards *d*, the same distance as between (7.0) and (9.0), to obtain a point (9.50) as far above (7.50) as (9.0) is above (7.0).

17. *Prob.* 3, *Fig.* 5. *Through a point in a plane to draw a line in the plane with a given inclination.*

Let *cd* be the scale of declivity of the given plane, and *a* (5.50) the given point; and suppose, for example, that the declivity of the plane is $\frac{1}{6}$ and that of the required line is $\frac{1}{10}$.

Draw the horizontal of the plane (5.50) which passes through the point, and any other horizontal, as (7.0). The projection of the required line will pass through *a*, and the portion of it between the two horizontals will be equal, *Art.* 6, to the difference of their references, or 1.5 ft. divided by the fraction which represents the inclination of the required line. Describing, therefore, from *a*, an arc, with this distance *ac* or $1.5 \div \frac{1}{10} = 15$ ft. as a radius, and joining the point b, where it cuts the horizontal (7.0), with a, this will be the projection of the required line.

18. *Prob.* 4, *Pl.* 1, *Fig.* 6. *Having three points of a plane, to construct its horizontals and scale of declivity.*

Let a (12.0), b (15.25), and c (15.50), be the projections of the three points. Join a with the other two, and construct the scales of declivity of the lines of junction, *Art.* 6. The lines joining the same references on these two scales will be horizontals of the required plane. Its scale of declivity is constructed by drawing two parallel lines perpendicular to the horizontals, and writing the references of the points where they intersect the horizontals.

19. *Prob.* 5, *Pl.* 1, *Fig.* 7. *To find the horizontals of a plane passed through a given line and parallel to another line.*

Let ab and cd be the projections of the two lines. From a point (10.0) on cd draw a line df, *Prob.* 2, parallel to ab; and by *Prob.* 4 find the horizontals of the plane of df and cd; these will be the required horizontals.

20. *Prob.* 6, *Pl.* 1, *Fig.* 8. *To find the horizontals of a plane the declivity of which is given, and which passes through a given line.*

Let bd be the scale of declivity of the given line, and suppose, for example, the declivity of the line to be $\frac{1}{15}$ and that of the required plane to be $\frac{1}{6}$.

Since the horizontals of the plane must pass through the points of the line having the like references, and as the distance in projection between any two of them, *Art.* 13, will be equal to the difference of their references divided by the fraction giving the declivity of the plane, it follows that to find the one drawn through b (14.0), for example, it will be simply necessary to describe from any other point, as a (12.0), an arc of a circle, with a radius of 12 ft., equal to the quotient just mentioned, and to draw a tangent to this arc from b. If any other horizontal, as (16.0), is required, which would not intersect the projection of the given line within the limits of the drawing; any two points, as (12.0) and (14.0), for example, may be taken as centres, and two arcs be described from them, with radii of 12 and 24 ft., calculated as above, and a line be drawn tangent to the arc; this tangent will be the required horizontal.

21. *Prob.* 7, *Pl.* 1, *Fig.* 9. *Having either the horizontals or the scales of declivity of two planes, to find their intersection.*

Join the points ab where any two horizontals, as (12.0) and (14.0), in one plane intersect the corresponding horizontals of the other, and the line so drawn will be the projection of the required intersection.

22. When the horizontals of the two planes are parallel, or when they are so nearly parallel that their points of intersection cannot be accurately found, the following method may be taken: Draw any two parallel lines as *cd*, *c'd'*, *Pl.* 1, *Fig.* 10; these may be considered as the horizontals of an arbitrary plane, and having the same references, (12.0) and (14.0), as the two corresponding horizontals in each of the given planes. The intersections of the horizontals of the arbitrary plane with those of the given planes will determine two lines, *mn*, *m'n'*, which, being the projections of the intersections of the given planes with the arbitrary plane, will, by their intersection *o*, determine the projection of a point common to the three planes, and therefore a point of the projection of the intersection of the two given planes. Assuming any other two parallels *ab*, *a'b'*, as the horizontals of another arbitrary plane; finding in like manner the point *o'* and joining *o* and *o'* by a line, this will be the required projection.

When the horizontals of the two planes are parallel, one point, as *o*, will be sufficient to determine the required projection, as it will be parallel to the horizontals.

23. *Prob.* 8, *Pl.* 1, *Fig.* 11.   *To find where a given line pierces a given plane.*

Through the projections of any two points of the given line, as *m'*, *n'*, having the same references, (12.0), (14.0), as two horizontals of the given plane, draw two parallel lines, *ab*, *a'b'*, which may be taken as the horizontals of an arbitrary plane. The projection of the line of intersection, *mn*, of this plane with the given plane being determined by *Prob.* 7, the point *o* where it intersects the projection of the line *m'n'* will be the projection of the required point, the reference of which can be found from the scale of the plane.

24. *Prob.* 9, *Pl.* 1, *Fig.* 12.   *To draw from a given point a perpendicular to a given plane, and find its length.*

Let *a* (12.0) be the projection of the given point; and let the given plane be represented by its scale of declivity.

The projection of the required perpendicular will pass through *a*, and be parallel to the scale of declivity of the given plane. The angle which it makes with the plane of reference is the complement of that between this plane and the given plane; its tangent therefore will be the reciprocal of the tangent of that of the given plane.

Drawing therefore through *a* the line *ac* parallel to *bd*, and constructing its scale of declivity, *Art.* 7, this will be the projection of the required perpendicular. The projection of the point *o* where it pierces the given plane is found by *Prob.* 8, and the true length of the perpendicular by *Art.* 8.

25. **Geometrical and Irregular Surfaces.**

All other surfaces may, like the plane, *Art.* 7, be represented by the projections on the plane of reference of the curves or lines cut from them by equidistant horizontal planes, together with the references of these curves; as many of these projections being drawn as may be requisite to determine all the points of the surface with accuracy; and their references being written in the same way as those of the horizontals of a plane.

In the more simple geometrical surfaces, a single horizontal curve, with the projection of some point or line of the surface, will alone suffice. For example, the cone may be represented by the projection and reference of any curve cut from it by a horizontal plane, with the projection and reference of its vertex; a cylinder by the projection and reference of a like curve, with the projection and reference of its axis, or of one of its right line elements; a sphere by the projection and reference of its centre and that of its great circle parallel to the plane of reference.

26. This method of projection is more particularly advantageous in the representation of irregular surfaces which, like the natural surfaces of ground, for example, are not submitted to any geometrical law, and in solving the various problems of tangent and secant planes to surfaces of this character. These surfaces can, for the most part, be alone represented by the projections of the horizontal curves cut from them by equidistant horizontal planes, and by supposing the zone of the real surface contained between any two horizontal curves to be replaced by an artificial zone, subjected to some geometrical law of generation, which shall give an approximation to the real surface sufficiently accurate for the object in view. The usual method of doing this is to take two consecutive horizontal curves as the directrices of the artificial surface of the zone, and to move a right line so as to continually intersect each of them, and be perpendicular to the consecutive tangents to one of them, the upper one being usually taken for this last condition.

If in *Pl.* 1, *Fig.* 13, for example, (6.0), (7.0), &c., are the projections of the horizontals of a surface, the zone between the curves (6.0) and (7.0) may be replaced by an artificial surface, the position of the projection of the generatrix of which, at any point of the upper curve (7.0), will be deter-

mined by constructing the horizontal tangent at that point,
as a, for example, and drawing ab perpendicular to it and
intersecting the lower curve. The position of the generatrix
$a'b'$ at any other point a' is constructed in like manner.

27. To obtain any horizontal of the artificial zone intermediate to the two directrices, it will be only necessary to
construct several positions of the generatrix, and to find on
these the points having the same reference as the required
curve. The horizontal of the surface (6.50), for example,
will bisect the projections of the generatrix in its various
positions.

**Problems of Irregular Surfaces and the Right
Line and Plane.**

28. *Prob.* 10, *Pl.* 1, *Fig.* 14.   *Through a given point in
a vertical plane which intersects a surface, to draw a tangent
to the curve of intersection of the plane and surface.*

Let a (5.50) be the given point, and ab the trace on the
plane of reference of the given plane. The points where
this trace intersects the horizontal curves of the surface will
be the projections of points of the curve cut from the surface
by the plane.

Let any arbitrary line as ac be now drawn through a,
and its scale of declivity be constructed; and let lines be
drawn between the points having the same references on ac
and on the horizontal curves where ab intersects them. These
lines will be the projections of horizontal lines and will gen-
erally make different angles with ac. The one as (7.0),
which makes the smallest angle with it, towards the descending portion, will determine the projection o of the tangential
point. For, construct the scale of declivity of the line of
which a (5.50) is the projection of one point, and o (7.0), on
ab, another. Comparing now the references of the points
on the line, and which is assumed as the projection of the
required tangent, with the references of the points of the
curve having the same projection, it will at once be evident
that these two lines have only the point projected in (7.0) in
common, and that every other point of the right line, of
which aob is the projection, is exterior to the curve, and
therefore the line itself must be tangent to the curve at the
point determined as above.

29. *Prob.* 11, *Pl.* 1, *Fig.* 15.   *To construct the elements
of a cone, with a given vertex, which shall envelope a given
surface.*

Let (10.0), &c., be the horizontals of the given surface;
and a (6.0) the projection of the vertex of the cone.

From a, draw lines ab, ab', &c., as the horizontal traces of vertical planes which pass through the vertex and inter sect the surface. Construct, by *Prob*. 10, the tangents from *a* to the curves cut from the surface by the planes *ab*, &c. These tangents will be the required elements.

30. *Prob*. 12, *Pl*. 1, *Fig*. 15. *To find the curve of intersection of a cone enveloping a given surface by a horizontal plane.*

Let (9.0) be the reference of the given horizontal plane. Having found, by *Probs*. 11 and 12, the elements of the cone, and constructed the scale of declivity of each one; then joining the points *o*, *o'*, *o''*, having the same reference on each scale as the given horizontal plane, a continuous line *mo''o'on* will be obtained, which will be the projection of the points where the elements pierce the given plane, and therefore the projection of the required intersection.

31. *Prob*. 13, *Pl*. 2, *Fig*. 1. *A limited extent of surface being given, and a point exterior to it, to find the limits within which planes may be passed through this point and lie above all the given surface.*

Let *a* (8.0) be the projection of the given point; (10.0), (9.0), &c., the horizontals of the given surface, the limits of which are the sector contained within the arc *BDC*, and the two radii *aB* and *aC*.

Taking *a* as the vertex of a cone which shall envelope the given surface, the elements of this cone can be found by *Probs*. 11 and 12. Any plane tangent to this cone, which does not intersect the surface within the given limits, will satisfy the conditions of the problem.

From the position of the vertex of the cone with respect to the surface, it will be seen that a horizontal plane, passed through the vertex, will cut from the cone two elements which will be projected in the two horizontals *ab'* and *ab''* (8.0) of the cone, the first of which will be tangent to the horizontal (8.0) of the surface, and the second *ab''* will pierce the surface, where the limiting arc *BDC* cuts the same horizontal (8.0); and that all the elements projected within the angles *Bab'* and *Cab''* will lie below the horizontal plane (8.0). Now, if the elements within these angles be prolonged beyond the vertex, they will form two portions of cones having the same elements as the portions below the vertex, and it is evident that any plane passed tangent to either lower portion, as *b'aB*, within one of these angles, will leave this portion below it, and the corresponding portion, formed by the prolonged elements, above it; and, in order that this plane shall satisfy the conditions of the problem, it must also leave the portions of the cone within the

angles $b'ab''$, and $b''aC$, also below it. The same reasoning
applies to planes passed tangent to the portions of the cone
within each of the other two angles. It is therefore evident
that a plane, which shall satisfy the conditions imposed,
must leave all that portion of the cone which lies above the
horizontal plane (8.0) through the vertex, below it, and all
the prolonged portions, corresponding to the portions below
the plane (8.0), above it.

To find any such plane, let the cone be intersected by a
horizontal plane, as (9.0), by *Prob.* 12. This plane will cut,
from the portion of the cone within the angle $b'ab''$, a curve
of which *non'* is the projection; the two extreme points of
this curve, within the limits, being at the points nn', where
the horizontal (9.0) of the surface cuts the limiting arc; it
will also cut, from each of the prolonged portions, a curve,
the one mr, and the other $m'r'$; the extreme point m of mr
being on the prolongation of the extreme element aC; that
m' of the other on the extreme element aB, on the other
side, prolonged. Having obtained these three curves, let
tangent lines, ms, $m's'$, be drawn, from the points m and m',
to the curve non'. A plane passed through either of these
tangents and through the corresponding element of the cone
as or as', drawn through the tangential point, will be a tangent plane to the cone; and as either of these planes will
leave the curve non' on one side of it, and the two curves
mr, and $m'r'$, on the other, it will leave all the portion of
the cone corresponding to the first curve below it, and the
portions corresponding to the other curves above it; and
will therefore satisfy the required conditions. The same will
hold true for any tangent plane to the cone along any element drawn between the points s and s'; since the tangent
drawn to any point of the curve non', between the points s
and s', will leave this curve on one side of it, and the other
two, mr and $m'r'$, on the other.

32. *Prob.* 14, *Pl.* 1, *Fig.* 16. *Through a given line to
pass a plane tangent to a surface.*

1st. Let ab be the projection of the given line, and (10.0),
(9.0), &c., the horizontals of the surface. From the points
on the line, as (10.0), &c., draw lines tangent to the horizontals having the same references; the tangent which makes
with the projection of the line the least angle towards the
descending portion, will, with the line, determine the required plane.

For, let the tangent (10.0) be the one which makes with
ab the least angle; from the other points, (9.0), &c., of ab,
draw lines parallel to the tangent (10.0); these lines will lie

in the plane that contains this tangent and *ab*, and will be horizontal of this plane; they also lie respectively in the planes of the horizontals (9.0), (8.0), &c., of the surface, but, since they fall exterior to these horizontals, it follows that their plane also lies exterior to every horizontal curve of the surface, except at the curve (10.0), and where it touches the surface at the point of contact of its horizontal (10.0) with this curve.

2d. When the line *ab*, *Pl.* 1, *Fig.* 17, is horizontal, let tangents be drawn to the horizontal curves and parallel to *ab*. These tangents may be regarded as the elements of a cylinder which envelops the surface, the tangent plane to which will be tangent to the surface. To find the element of contact of the plane and cylinder, let the cylinder and given line be intersected by an arbitrary vertical plane, of which *od* is the trace. From the point *o*, (6.5), where the line pierces this plane, let a tangent line be drawn to the curve cut from the cylinder by the plane, by *Prob.* 10. The point of contact will determine the position of the element of the cylinder along which the plane, through *ab*, will be tangent; since the tangent to the curve projected in *od*, with the line *ab*, will determine the tangent plane to the cylinder.

3d. When the line *ab*, *Pl.* 1, *Fig.* 18, is so nearly horizontal that tangents cannot be drawn from its points, within the limits of the drawing, to the horizontal curves. Let any point of the line, as *o*, (7.0), be taken as the vertex of a cone enveloping the surface; a plane passed through the line and tangent to the cone will be tangent to the surface.

Find, by *Probs.* 10 and 11, the projection *mrn* of the curve cut from this cone by the horizontal plane (8.0); from the point (8.0) of *ab* draw a tangent to *mrn*. This tangent, with the line *ab*, will determine the required plane.

33. *Prob.* 15, *Pl.* 1, *Fig.* 19. *To find approximately the point where a given right line pierces a surface.*

Let (8.0), (9.0), &c., be the horizontals of the surface, and *df* the scale of declivity of the line. Through any two points, as *a* (9.0) and *c* (8.0), draw two parallel lines, as *am* and *cn*, which may be taken as the horizontals of an arbitrary plane passed through the given line. Joining the points *m*, *n* where the horizontals of the arbitrary plane intersect the corresponding horizontals of the surface, this line *mn* will be the approximate intersection of the plane with the zone of the surface between the horizontals (8.0) and (9.0), and the point *o* where *mn* intersects *df* will be the approximate point required.

34. *Prob.* 16, *Pl.* 1, *Fig.* 20.   *To find the intersection of a plane and surface.*

Let (10.0), (9.0), &c., be the horizontals of the surface, *ef* the scale of declivity of the plane.

Draw the horizontals of the plane having the same references as the horizontals of the surface, the points of intersection of the corresponding lines will be the projections of points of the required intersection.

When it is desired to find a point of the curve of intersection intermediate to two horizontal curves; if the reference of the required point is fixed, it will be necessary to construct, *Art.* 27, the horizontal of the surface, and the horizontal of the plane having this reference; their intersection will give the projection of the required point. If the reference of the required point is not fixed, draw any generatrix, as *ac* of the zone on which the required point is to be found, and by *Prob.* 8, *Fig.* 11, find the projection of the point, as *o*, where *ac* pierces the given plane; this will be the required point.

35. **Application of Preceding Problems.**

The following problems will aid as illustrations of the preceding subject in its application to the determination and delineation of lines and surfaces.

36. *Prob.* 1, *Pl.* 2, *Fig.* 2.   *The plane of site of a work, the exterior line and scale of declivity of its terre-plein being given; to construct the plane of the rampart-slope and its foot; also a ramp of a given inclination along the rampart-slope leading from the plane of site to the terre-plein.*

Let *a* (74.50) and *b* (76.0) be the references of two points on the exterior line of the terre-plein, and *mn* its scale of declivity; let the rampart-slope be $\frac{2}{3}$, the declivity of the ramp $\frac{1}{9}$, its width 4.30 yards; and the plane of site be horizontal and at the ref. (60.0).

The foot of the rampart-slope lying in the plane of site will be horizontal, and will be determined, *Prob.* 6, *Fig.* 8, by finding the line of the slope at the ref. (60.0).

Having the two bounding lines of the rampart-slope, the inner line *cd* of the ramp is constructed, by assuming a point *c*, on the foot of the rampart-slope, as the point of departure, and determining the line of $\frac{1}{9}$ drawn from *c* on the rampart-slope by *Prob.* 3, *Fig.* 5. Having found this line, which is also the line of greatest declivity of the ramp, the exterior line *ef* of the ramp is drawn parallel to it, and at a distance 4.30 yds., equal to the width assumed for the ramp. The horizontals of the ramp will be perpendicular to these two lines. The foot of the ramp, *ce*, will be a hor-

izontal line, drawn through the point of departure. The
top of it, *df*, will be determined by *Prob.* 7, *Fig.* 9, by finding the intersection of the ramp and the terre-plein, one
point of which will be the point *d* (76.30), the intersection
of the inner line of the ramp and the interior line of the
terre-plein.

The ramp is terminated on the exterior, by passing a
plane through its exterior line *ef* having the same slope as
the rampart-slope. This plane will intersect the plane of
site in a line parallel to the foot of the rampart-slope, and
the terre-plein in one parallel to the exterior line of the
terre-plein.

37. *Prob.* 18, *Pl.* 2, *Fig.* 3. *Having given the lines of
the parapet of a work and the scales of declivity of the planes
of its interior crest and terre-plein, to determine the lines and
surfaces of a barvette in its salient for five guns.*

Let *ab* be the scale of declivity of the plane of the interior
crest, which, as the terre-plein is parallel to the plane of
the interior crest and 8 feet below it estimated vertically,
will also serve as the scale of declivity of the terre-plein, by
subtracting 8 feet from the references of the former to obtain
the corresponding references of the latter. Having constructed a pancoupé of 4 yds. in the salient, find the intersection of the top surface of the barbette, which is horizontal
and assumed on the drawing at the reference (82.75) with
the planes of the interior slope, this intersection will determine the foot of the genouillère of the barbette. From this
last line at the pancoupé set back along the capital a distance
of 8 yds., and from the extremity of this line draw a perpendicular to the interior crest of each face. The pentagonal figure thus marked out will be the space for the gun in
the salient. From the foot of each of the perpendiculars
set off along the faces distances of 12 yds. for the lengths
along the interior crests to be occupied by two guns on each
side of the salient. Setting back from the extremities of
these two last distances perpendiculars to the interior
crest of 8 yds. and drawing lines through the extremities of
these perpendiculars parallel to the interior crests, they with
the two perpendiculars will mark out the exterior bounding
lines of the barbette. By passing planes of ¼ or 45° through
these exterior lines, and finding by *Prob.* 7, *Fig.* 9, their intersections with the terre-plein, these lines will be the foot
of the barbette slopes. A ramp having a slope ⅛ leads from
the terre-plein to the top of the barbette; the width of this
ramp is 3.30 yds., its interior line in projection being on the
prolongation of the foot of the banquette slope. The ramp

is terminated by side slopes of ¼, the intersections of which with the terre-plein and the slopes of the barbette and ban‧quette are found by *Prob.* 7, *Fig.* 9. The foot of the ramp or its intersection with the terre-plein is also found by the same problem.

As the top surface of the barbette is horizontal, it may be necessary in some cases to make the interior crest along the barbette also horizontal, in which case the superior slope of the parapet along the barbette being higher than the rest of it, the two planes will be connected by a plane of 45°, as at *C*.

3S. *Prob.* 19, *Pl.* 2, *Fig.* 4. *To determine the bounding surfaces of a ramp leading up an irregular surface and so placed that its axis or centre line shall nearly coincide with the irregular surface.*

Let (8.0), (9.0), &c., be the horizontal curves of the sur‧face, and let *a* (8.0) be the point of departure or foot of the ramp. Assuming the declivity of the ramp ⅛, for example, from *a*, with a radius of 9 units, describe an arc, and join by a right line the point *b* where it cuts the horizontal (9.0) with the point. Repeat this construction from *b* to *c* on the horizontal (10.0); and so on to the top *e* or point of arrival. The broken line *a-b-c-d-e* will be the projection of the axis. But, to avoid the angular changes of direction, the straight portions of the axis may be connected at the angular points, by setting off from *b*, for example, the equal distances *ba′*, *bc′*, and connecting these points by an arc of a circle tangent to the straight portions. The same construction being re‧peated at the other angular points, the broken line will be replaced by the sinuous line *aa′c′*, &c., as the axis. Having determined the axis, the exterior and interior lines of the top surface are drawn parallel to the axis, and at a distance from it equal to half the assumed width of the ramp.

From the position of the axis the exterior half of the ramp will be in embankment and the interior in excavation. To determine the side slopes of the embankment pass planes through the straight portions of the exterior edge of the ramp, and find by *Prob.* 6, *Pl.* 1, *Fig.* 8, the horizontals of these planes, and by *Prob.* 16, *Fig.* 20, the intersections of these planes with the irregular surface. The plane surfaces of the side slopes thus determined are connected by curved sur‧faces which pass through the curved lines of the exterior edge. These surfaces may be determined as follows: Take, for ex‧ample, the point *n* at the foot of the plane side slope *A* where it cuts the radius through *a′* prolonged, of the arc *a′c′*; and the point *o* on the radius through *c′* where it cuts the foot of the plane side slope B. The lines of which nv and ou are the projections will evidently have the same inclination, and they may be assumed as the lines of junction of the plane slopes A and B and the curved side slope x. This curved side slope may then be generated by the motion of a right line which has the top line of which vu is the projection for its directrix, whilst in its motion it makes a constant angle with the plane of comparison, and its projections are constantly normal to the arc vu. From the construction comprising these conditions the foot no of the curved portion x of the side slope is determined. The same constructions are repeated to obtain the portions C of the plane, and y, z of the curved side slopes, with the line m-n-o-p-q-r-s the foot of these slopes.

The side slopes of the part in excavation A', B', C' and x', y' with the line m'-n'-o'-p'-q'-r' are determined by like constructions.

The portions of the top surfaces of the ramps bounded by the arcs of circles are helicoidal surfaces, of which the axis is the directrix and the plane of comparison the plane director.

The curved surface side slopes are also evidently helicoidal surfaces, the directrices of which are the curved lines above mentioned, and the vertical lines through the centres of the arcs which are the projections of those curved lines.

Remarks. In the figure the declivity of the side slopes of the embankment is one-half the excavation. The declivities of the curved portions of the top are greater than those of the plane surfaces, the difference depending on the angle between the straight portions of the axis.

39. **Observations on the best mode of executing Drawings.**

Accuracy. The first requisite in all drawings is *minute accuracy*, both in the geometrical constructions, and in writing down all letters and numbers which serve either as references, or to give dimensions. To attain this, so far as regards the geometrical part, judgment is to be exercised in the selection of the means for establishing on the drawing the positions of the various points which are either given or to be found; as one method although in theory as correct as some other may not, in practice, be found to yield as satisfactory results. The following remarks will serve to illustrate this point:

1st. *In setting off from a scale of equal parts several distances, along a line,* whether equal or unequal, the most accurate method is to commence by first setting off the entire

2

distance, and then the several parts; taking care to verify,
from the scale, the aggregate of the several partial distances;
thus in the example *Pl.* 2, *Fig.* 5, where the aggregate of
all the partial distances is 60.33 feet, commence by setting
off the entire distance 60.33 feet; next 50.33, which is the
sum of the two distances 20′ and 30′.33, then verify the re
maining 10′ by the scale.

2d. *When a distance to be set off is so small* that it cannot be laid down with accuracy by the points of the dividers,
the following method may be employed : set back, from the
point from which the required distance is to be set off, any
arbitrary distance, then set forward, from this last point, a
distance equal to the sum of this arbitrary distance and the
one required; thus in *Pl.* 2, *Fig.* 6, where 2′ is to be set off
from *a* towards *c*, set back from *a* say 30′ to *b*, then from *b*
32′ to *c*.

3d. *To set off a point at a given perpendicular distance
from a line,* it will mostly be found more speedy, and more
accurate, to take off from the scale the given distance, in
the dividers, and, setting one point on the paper, bring the
other so that the arc described by it, with the given distance
as a radius, shall be tangent to the line, than to employ the
usual method of first erecting a perpendicular to the line
and then setting off the required point along the perpendicular; thus in *Pl.* 2, *Fig.* 7, wishing to set off *c* at 20′ from
ab, take 20′ in the dividers, and, by the eye, find where one
point must be placed so that the other describing an arc will
touch *ab*.   This method will be found convenient in drawing
a parallel to a line at a given distance from it by setting off
another point in the same way, and drawing through the
two the required parallel.

4th. In setting off several points for the purpose of *drawing several parallels to a given line,* as, for example, the parallel lines which bound the planes of a parapet, it will be
found most speedy and accurate to draw first upon a slip of
smooth thin paper two lines perpendicular to each other,
then marking on one of the lines the respective given distances of the parallels from the other, and cutting the paper
close to the line along which the given points are marked
off, so that the strip when laid upon the drawing so as to
have one of its lines to coincide with that to which the parallels are to be drawn, their distances from it can be pricked
off by a sharp pointed pencil, or in any other way.   In *Pl.*
2, *Fig.* 8, *ab* is the line of the drawing ; *A* the strip of paper,
fc, *fd*, *fe*, &c., the distances at which the parallels are to be
drawn from *ab*, marked off on the edge of *A* perpendicular

to the line f, which line when A is laid on the drawing, should coincide with ab. If the line ab is somewhat long, it will be better to set off these points near each of its extremities and join them by lines, than to draw the parallels in the usual way by aid of the ruler and triangle.

5th. *When a point is to be constructed by means of the intersection of two lines arbitrarily chosen*, such a position should be assumed for the arbitrary lines that they shall not form a very acute angle at their point of intersection, as in that case this point might not be so distinct to the eye as to be marked with accuracy. For example, in erecting a perpendicular to a line at a given point, and in like problems, in which points are found by the intersections of arcs of circles, it will be best and most convenient to take for the radii of the arcs the distance between their centres, as the angle between the tangents to the arcs at their point of intersection will then be 60°, which is a sufficient angle to give accurately the point where the lines cross. In cases like *Figs.* 10, 11, *Arts.* 22, 23, the arbitrary lines ab, $a'b'$, &c., should be so chosen as to intersect the horizontals nearly at right angles, and so, also, that the resulting lines, by which the points o, o', are determined, shall not intersect in too acute an angle.

In all such cases of determining points, and where a point is pricked into the paper, it will be found well to *designate the point thus* ⊙, by a small circle drawn around it with the lead pencil, in order that the eye may see it with more distinctness.

6th. *In determining a portion of a line by the construction of two arbitrary points*, the points should be so chosen that the portion required may fall between them and not beyond them. In *Pl.* 1, *Fig.* 10, for example, if the required portion of the line of intersection of the planes extended on either side, beyond o, or o', or beyond both, the lines ab, cd, &c., should be so chosen as to bring o and o', as far apart, at least, as the length of the required portion of the line which they serve to determine.

7th. No means of verifying the accuracy of the construction of points, or lines, should be omitted. In *Pl.* 1, *Fig.* 9, for example, other corresponding horizontals should be drawn, and, if the line of intersection determined by the two points first found is correct, their points of intersection also will fall upon it. In *Pl.* 1, *Figs.* 9, 10, the scale of declivity of the line of intersection being determined, the references of the points, where it intersects the scales of declivity of the planes, should be the same as the same points on the scales, if the line has been accurately determined. A general and minute verification of all the parts of the drawing should be made before any portion of it is put in ink.

Neatness. This is a not unimportant element in the attainment of accuracy in drawing. A few minutiae, when attended to, will subserve this end.

That part of the paper on which the draughtsman is not working should be kept covered with clean paper, pasted on the edge of the board, so as to fold over the drawing, and the parts which are finished should be similarly protected.

Before commencing the daily work the paper should be carefully dusted, and the scales, rules and triangles be carefully wiped with a clean dry rag.

As few lines of construction as possible should be drawn in pencil; and only that part of each which may be strictly necessary to determine the point sought. As, for example, where a point is to be found by the intersection of two arcs of circles; when the position of the point can be approximately judged of by the eye, only a portion of one arc, which will embrace the point, may be drawn, and the point where the second arc would intersect the first be marked without describing the arc. In *Pl.* 1, *Fig.* 10, instead of drawing the entire lines *ab, cd.* &c., it would be simply necessary to mark the points only where they cut the horizontals; and, in like manner, the points *o* and *o'* might be marked without drawing the entire lines.

No more of any line of the drawing should be made in pencil than what is to remain permanently in ink. The object of these precautions is to keep the paper from becoming covered with dirt and the lines from being defaced by the wear of the paper.

Inking. In inking the lines the following directions will be found useful:

Efface carefully all pencil lines that are not to be inked; and those parts of the permanent lines which are not to remain, before commencing to ink.

When right lines are tangent to curves, put in ink the curve before the right line; draw all arcs of equal radii at once, one after the other; if several arcs are to be described from the same centre, it will be well to put a thin bit of quill over the point for the end of the dividers to rest on, to avoid making a large hole in the drawing.

If the drawing is not to be colored with the brush, all the lines of one color should be put in before commencing on those of another.

If one of the bounding lines of a surface is to be made heavier than the others, its breadth should be taken from

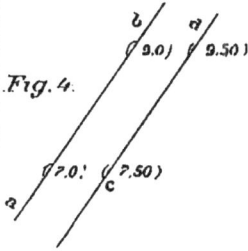

Fig. 4.

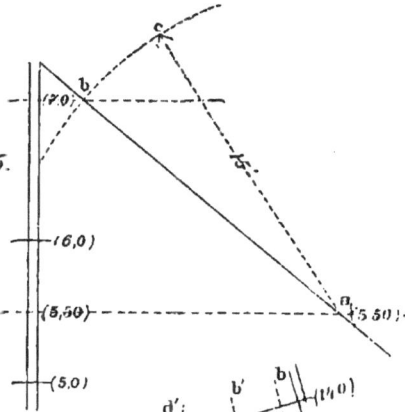

Fig. 5.

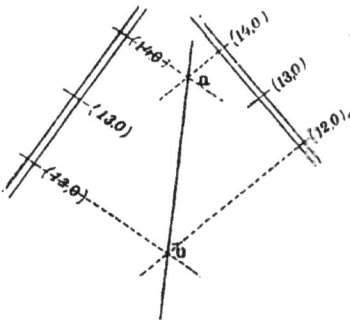

Fig. 9.

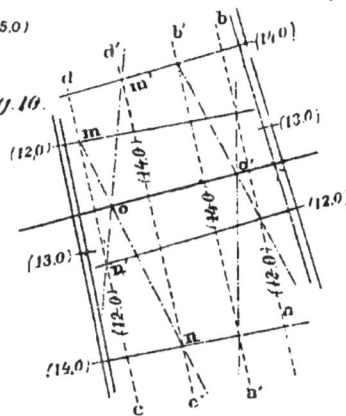

Fig. 10.

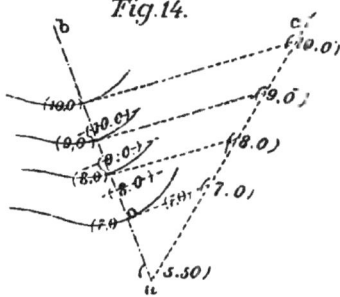

Fig. 14.

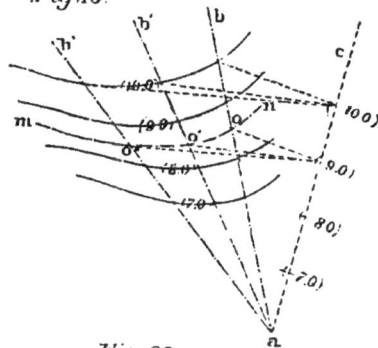

Fig. 15.

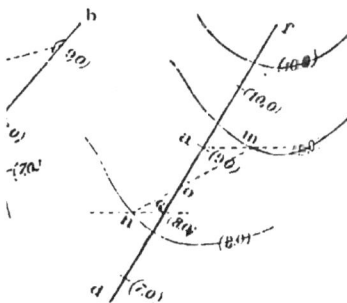

Fig. 19.

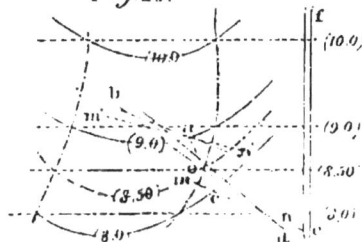

Fig. 20.

Plate 1

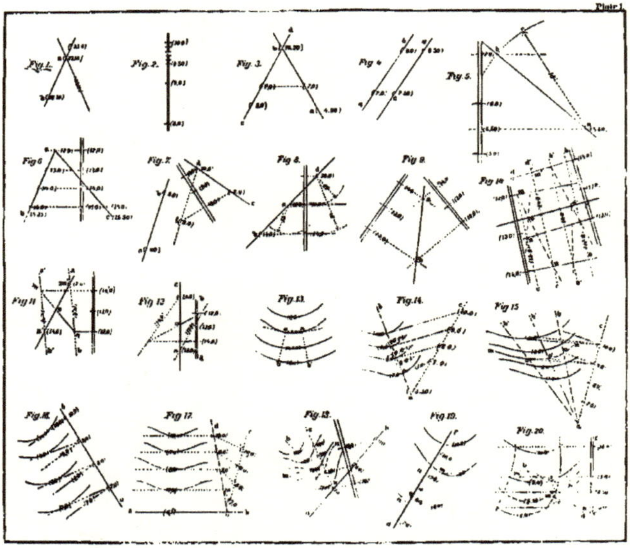
Fig. 1.
Fig. 2.
Fig. 3.
Fig. 4
Fig. 5.
Fig. 6
Fig. 7.
Fig. 8.
Fig. 9.
Fig. 10
Fig. 11
Fig. 12
Fig. 13.
Fig. 14.
Fig. 15
Fig. 16.
Fig. 17.
Fig. 18.
Fig. 19.
Fig. 20.

the surface they limit and not be added to it; ard when the heavy line forms the boundary of two surfaces, its breadth must be taken from the one of greatest declivity.

Coloring. When the drawing is to be colored, all lines that are not to be black may be put in first with black, making them very faint, so that they may receive their appropriate colors after the drawing is otherwise completed.

No heavy line should be put in until the work with the brush is completed.

When all the lines are in, the drawing should be thoroughly cleaned with stale bread-crumb; and then have several pitchers of water dashed over it, the board being placed in an inclined position to allow the water, colored by the ink lines, to escape rapidly, and not to discolor the paper.

In using the brush, whether for flat tints, or graded, the requisite depth of tint should be reached by a number of faint tints laid over each other; this is especially necessary in laying tints of blacks, browns, and reds.

To obtain an even flat, or graded tint, on dry paper requires considerable skill. The best plan for this is, first to wet with a large brush, or clean rag, the surface on which the tint is to be laid, then, with a slightly moist rag, clear the surface of water, and before the paper has time to dry to lay on the tint. With this precaution, the heaviest tints of Chinese ink, the most difficult of all to manage on dry paper, can be neatly laid down.

Titles, &c. The lettering and numbering of a drawing should be in ordinary printed character; this is particularly requisite in the numbering, to avoid misapprehensions which often arise from individual peculiarities in writing numbers.

As has been already remarked, references are written in black, within brackets which, when practicable, embrace the point referred to. When not practicable, a small dotted line may lead from the point to the reference; thus, $\odot$...(25.50); but to distinguish references from other numbers the designation of the unit is omitted.

All horizontal distances between points are written upon a dotted line drawn between the points, with an arrow-head at each end; where several partial distances in a right line are marked, it will be also well to mark the total distance: the latter may be written above or beneath the former, *Pl. 2, Fig. 5.*

In writing horizontal distances, the usual designation of the unit is always written thus, y for yards, $'$ for feet, &c. All the numbers must be expressed in the same unit; the fractional parts being in decimals.

References and horizontal distances cannot be too much multiplied, in order to avoid misapprehensions, and the results of errors of construction, as well as to save the time that would be taken in applying dividers to the drawing to find from the scale affixed to it the dimensions of any part.

Scale. A scale very accurately constructed should be affixed to the drawing before it is cut from the board; so that the shrinkage of the paper, which is about $\frac{1}{300}$, may affect all the parts equally, and the scale thus be made to correspond to the real lengths of the lines on the drawing. The scale should be divided according to the decimal system, as being most convenient for counting off.

The first division of the scale should furnish the units and also their decimal parts, if the scale bears that proportion to the true dimensions of the object represented which will admit of these divisions. This first division is numbered from right to left, *Pl.* 2, *Fig.* 9, the zero point being on the right, the 10 point on the left; the succeeding divisions, to 50 inclusive, should each be equal to the first division, containing ten units each. The remaining divisions may contain fifty units each. It will be seen that any number of tens, units, or fractional parts of a unit can thus be readily taken off from the scale by the dividers. The scale should be long enough to give the dimensions of the longest line on the drawing.

The proportion which the scale bears to the true dimensions of the object should be written above the scale; thus, *Scale one inch to ten yards, or* $\frac{1}{300}$. And the designation of the unit of the drawing should be annexed to the last division on the scale, as *yds.* for yards, *ft.* for feet, &c.

Note.—For more detailed directions on the mechanical or instrumental methods of geometrical drawing, see Mahan's *Industrial Drawing.*

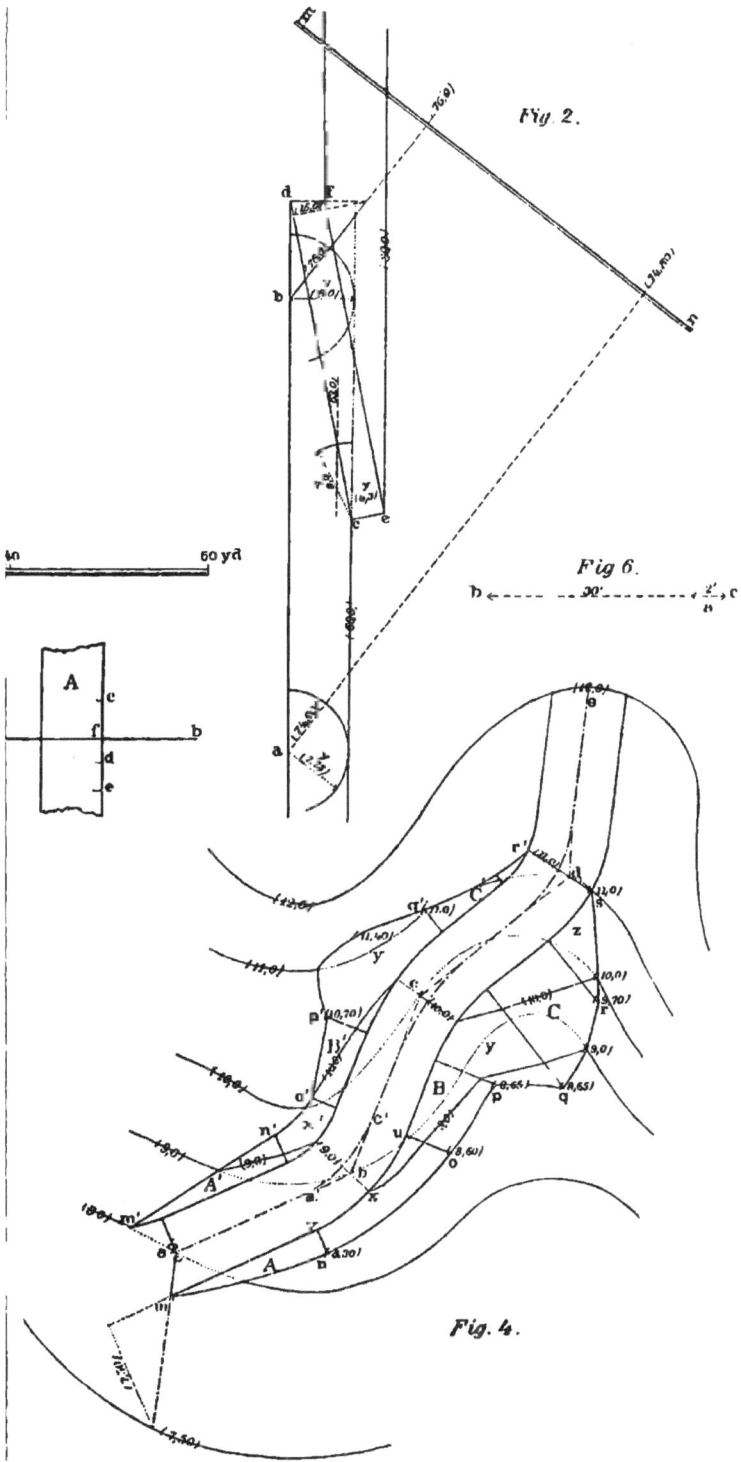

Fig. 2.

Fig 6.

Fig. 4.

50 yd

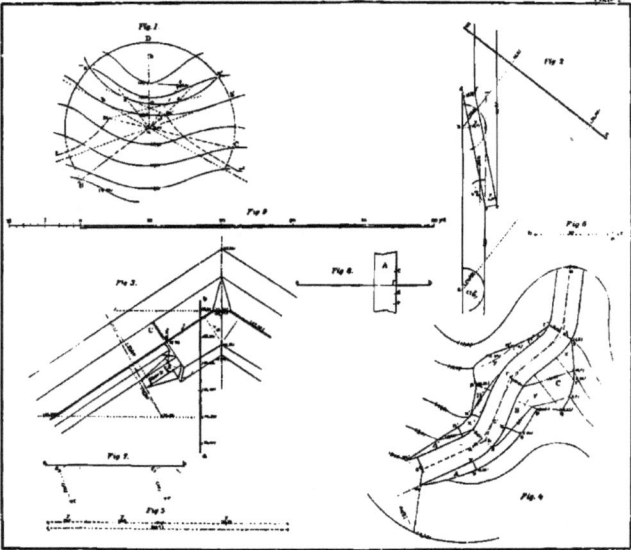

Plate 2

STONE CUTTING.

The object of this Article is to explain the geometrica methods of representing the more usual and elementary combinations of blocks of stone in walls and arches, by means of their projections; and from these and the data of the problem to deduce the true dimensions of the bounding surfaces and lines of each part.

Walls bounded by Plane Surfaces. In walls of cut stone[*] the blocks are usually separated by horizontal and vertical joints; the latter being in vertical planes perpendicular to the face of the walls, and which are termed *planes of right section*, to distinguish them from other vertical planes of section. When the face of the wall is inclined to the horizon, its slope, or *batir*, is usually expressed by the ratio of the base of the slope to the perpendicular, measured in the plane of right section; or the slope is said to be so many base to so many perpendicular. In the right section $A'B'C'D'$, (*Pl. A, Fig.* 1,) for example, the inclination of the face $A'C''$ to the base of the wall $A'B'$, is measured by dividing the perpendicular $C'E'$ from C' upon the line $A'B'$, by the distance $A'E'$ between the point A' and the foot of the perpendicular. The quotient $\frac{C'E'}{A'E'}$ thus obtained, is evidently the natural tangent of the angle $C'A'E'$; and the most convenient method of representing the batir is by a fraction; the numerator expressing the number of units in the perpendicular, and the denominator the corresponding number of units in the base; thus a batir of $\frac{6}{1}$ expresses a slope of six perpendicular to one base; a batir of $\frac{3}{2}$, one of three perpendicular to two base, &c.

Prob. 1. *Having given the right section of a wall, to construct the projections of its bounding lines, and the edges of the horizontal and vertical joints.*

Let $A'B'D'C'$, (*Pl. A, Fig.* 1,) be the right section; the base $A'B'$ and the top $C'D'$ being horizontal; the face $A'C'$ having a batir $\frac{C'E'}{A'E'} = \frac{6}{1}$; and the back $B'D'$ vertical.

Draw a line $AA_{,}$, to represent the foot of the face in plan. Parallel to $AA_{,}$ draw $CC_{,}$, and at a distance from it equal to

* See Mahan's *Civil Engineering*, Art. 351.

$A'E'$, the distance of the point A' from the foot of the perpendicular drawn from C'' in the plane of right section; $CC_,$ will be the top line of the face in plan. Parallel to $CC_,$, and at a distance $C'D$, the breadth of the wall at top, draw $BB_,$, which will be the projection of the back.

At any convenient distance from $BB_,$ draw $A''A'''$ parallel to it, to represent the foot of the wall in elevation; the top $C''D''$ will be drawn parallel to $A''A'''$, and at the height $B'D'$ of the top above the base.

To draw the projections of the horizontal edges, suppose the wall divided into four equal courses by horizontal joints. As the batir of the wall is $\frac{6}{7}$, the base of the slope of each of these equal courses will be one-sixth of its height; if lines therefore are drawn parallel to the foot $AA_,$, and at a distance from each other equal to $\frac{1}{6}$ the height of each course, these lines will be the projections in plan of the horizontal edges, as shown on the right of the plan. As the projections of the vertical edges are contained in planes of right section, they will be drawn perpendicular to the horizontal ones, and breaking joints with them; as represented on the same portion of the plan.

The horizontal edges in elevation will be drawn parallel to $A''A'''$, and at a distance from each other equal to the height of a course. The vertical edges will be drawn perpendicular to these last, and corresponding to their projections in plan.

Remark. If the projections of any horizontal line of the face, at a given height above the foot of the wall, are required, as mn and $m''n'$, for example, it is evident that $m''n'$ will be drawn in elevation at the given height above the foot $A''A'''$, and that mn will be parallel to $AA_,$ in plan, and at a distance from it equal to $\frac{1}{6}$ the height of $m''n'$.

Prob. 2. *Having given the batir of the faces of two walls that intersect, the foot of each being in the same horizontal plane, to draw the projections of the line of intersection of the faces.*

Let Aa, (*Fig.* 1,) be the foot of one wall in plan, and the batir of its face $\frac{6}{7}$; ab the foot of the other, and the batir of its face $\frac{4}{7}$.

It is evident that the point a will be one point of the intersection in plan. If now a horizontal line be drawn in each face at the same altitude, they will intersect and give a second point of the intersection of the faces. Assuming any altitude for the horizontal line on one face, its projection mn, in plan, will be parallel to Aa, and at $\frac{1}{6}$ of the assumed altitude from it. In like manner, the projection *no* of the corresponding line on the other face having the batir $\frac{4}{7}$ will be parallel to ab, and at $\frac{1}{4}$ the assumed altitude from it. Drawing, therefore, a line through the points a and n, this will be the intersection in plan.

To obtain the intersection in elevation, draw the foot of the wall $A''a'$ in elevation, also the line $m''n'$ at the assumed altitude. The point a will be projected in a', and the point n in n'; and the line $a'n'$, drawn through them, will be the intersection in elevation.

Prob. 3. *To construct the projections of the bounding lines and edges of the joints of a buttress against the inclined face of a given wall; the base of the buttress being given, and being in the same horizontal plane as the base of the wall; the faces of the buttress, its end, and the top to have given slopes.*

Along the foot AA, set off the breadth of the buttress ad at its base, and construct the two sides ab, cd; and the end bc of the base. Let the batir of the face of the wall be $\frac{9}{7}$; that of the two faces and the end of the buttress $\frac{4}{7}$; and that of its top $\frac{1}{6}$.

By *Prob.* 2 construct the projections, in plan and elevation, of the intersections ae, $a'e'$, and dh, $d'h'$, of the faces of the buttress and wall; and those bf, $b'f'$, and cg, $c'g'$, of the end and faces.

To construct the projections of the top surface of the buttress.

Suppose that the top line of the buttress eh, $e'h'$, where it joins the wall, is of the same altitude as the top of the wall; and that the top surface from this line outwards from the wall has the given slope $\frac{1}{6}$. Now, if a line as $y'z'$ be drawn parallel to $e'h'$ the top line, and at any assumed distance below it, this line may be regarded as the projection of a horizontal line in the top surface of the buttress; and its corresponding projection in plan will be a line yz parallel to eh, and at six times the distance from it that $y'z'$ is below $e'h'$. Having drawn these two lines of indefinite length, construct the projections of the horizontal in the face of the buttress which is at the same distance below the top. The projection in elevation will be a continuation of the same line $y'z'$, and in plan its projection will be parallel to the foot ab, and at a distance from it equal to $\frac{1}{4}$ its altitude above it. Drawing an indefinite line xy parallel to ab at this distance from it, the point y, where it intersects the line yz, will be a point of the projection in plan of the intersection of the top surface and face of the buttress. The point e is another point; joining e and y by a line and prolonging it unt.l it intersects the projection of the intersection of tho face and end at f, this line ef will be the projection of one side of the top surface in plan. The other side hzg will be found by a like construction. The points f and g being joined will be the projection of the end of the top surface.

To obtain the corresponding lines in elevation; the points y and z are projected into y' and z'; the points e'. y', and h', z', are joined by lines which are prolonged to meet the lines which are the projections of the exterior edges of the buttress at f' and g', which correspond to f and g in plan, and the points f' and g' are joined; $e'f'g'h'$ is the elevation of the top surface.

The projections of the edges of the vertical joints in plan will be perpendicular respectively to the lines bc and cd, and breaking joints as shown on the right portion of the plan. The projections of these lines in elevation will be found by projecting their extremities into the corresponding projections of the horizontal edges in elevation, as shown on the elevation of the face and a portion of the end, on the right.

Prob. 3, *Case* 2 (Pl. B, Figs. 1, 2, 3). *Having given the cross section of a brook, or other small natural water way, over which a full centre arched stone culvert is to be thrown, to support an embankment of a roadway, of a given height above the natural surface of the ground, to construct the bounding lines of a wing wall with plane bounding surfaces.*

Figs. 1, 2 are the elevation and plan, or the vertical and horizontal projections of the parts; P Q being the ground line. Fig. 3 is a section and elevation, through the axis of the arch, on the vertical plane of which R S is the ground line. M, M′ are the slopes of the embankment. N, N′ the bottom of the brook.

Let C B Z′, Fig. 1, be the cross section of the side bank of the water way, and of the adjacent level ground : O the centre of the semicircle of the full centre arch, taken on the level C B of the natural surface; L I the level of the top of the embankment; L′ E′, Fig. 2, the foot of the embankment

The wing wall and arch (*Figs.* 1, 2), are supported upon a general substructure, the height of which is A A′; the plan of that portion of which, supporting the wing wall, is shown by A′ B′ C′. The faces of this substructure being vertical, and projecting a distance, represented by Z A, beyond the springing line of the arch, the foot of the wing wall, and the foot of the embankment; the point B′ being taken on the crest B′ K of the side bank.

Through the line Z D, Z′ D′, the foot of the wing wall,

the plane of its face is passed; the top point of which X, X′ is found by drawing a line *m n* parallel to Z′ D′, and at a distance from it equal to the base of the batir corresponding to the height of the embankment L I above D Z, and taking its intersection X, X′ with the top line of the head of the arch. Joining X Z, it will be the elevation of the intersection of the face of the wing wall with the end of the arch.

The top surface of the wing wall X″ D″ (*Fig.* 3), receives the same slope as the side slope of the embankment M, M′, and is here taken to coincide with it. The wing wall is terminated at the end by a vertical plane D″ F″ parallel to the head wall of the arch. The thickness I X, I′ X′ of the wing wall at top is assumed.

Joining the points X′ D′ and X D, the interior edge of the top line of the wing wall is found. Drawing I H and I′ H′ respectively parallel to these, the exterior lines are found.

The lower end of the wing wall is terminated by what is termed a *newel stone*, which serves, in this case, as a buttress. The height of this stone D″ F‴ is arbitrary, as is also its slope F‴ G″ on top. Assuming these, the intersection of the vertical plane, terminating the wing wall with its face, will be the lines F′ D′, F D, parallel respectively to X′ Z′ and X Z. The lines F′ G′, F G, and H′ E′, H E, which also are parallel, will be found by *Prob.* 3.

The vertical joints of the face of the wing wall are perpendicular to its face. Drawing the line X′ Y′ perpendicular to D′ Z′, and its corresponding projections X Y, X″ Y″ on *Figs.* 1, 3, the directions of the edges of the vertical joints, as *x′ y′, x y, x″ y″*, will be parallel, on their respective *Figs.*, to these lines.

The top of the wing wall, instead of a coping, is formed with elbow joints uniting with the horizontal joints. The portion of the joint forming the elbow is perpendicular to the top surface. Drawing then a line Z″ W″ (*Fig.* 3), perpendicular to D″ X″, and its corresponding projections Z W, Z′ W′ on Figs. 1, 2, these will be the directions z w, z′ w′, z″ w″, of the elbows. The depth of the elbow is arbitrarily assumed, by drawing a line p″ q″ on Fig. 3 parallel to D″ X″.

CYLINDRICAL

AND OTHER

ARCHES.

———————

To facilitate the geometrical operations for determining the bounding surfaces and lines of the voussoirs of arches, a few preliminary problems and theorems, on which these operations are based, will first be explained.

Prob. 4, (*Pl. A, Fig.* 2.) *Having given a semi cylinder, the right section of which is a semicircle, and its axis and two bounding elements being horizontal, to construct the projections of the intersection of the cylinder by a plane inclined to its axis and having a given inclination to the horizontal plane containing the axis; also, the projection of the intersection of this semi cylinder with another semi cylinder with a semicircle also for its right section, the axis and bounding elements of this last being in the same horizontal plane as those of the first; and then to develop the portion of the first semi cylinder which lies between the given plane and the other cylinder.*

Let $a'c'b'$ be the right section of the given cylinder, and o' its centre; the line $a'b'$ being horizontal. Let $a'A$ and $b'B$ be the horizontal projections of its bounding elements, and $o'C$ that of its axis. Let ab be the trace of the given inclined plane on the horizontal plane of the bounding elements; AB one of the bounding elements of the other cylinder, and LM its axis. The quadrant AL' the half of the right section of this cylinder; L the centre of this quadrant.

1st. Taking any two elements of the given cylinder, at the same height, as $x'x_2$ and $y'y_2$, above $a'b'$, they will be projected in plan parallel to the axis $o'C$, and will be drawn indefinitely through the points x_2 and y_2. The given inclined plane will cut these elements at the same height $x'x_2$, and if the projection xy of a horizontal line in this plane, at the height $x'x_2$, be drawn, the points x and y, where it cuts the two elements of the cylinder, will be two points of the required projection in plan. To construct this line xy, let the given inclination of the plane be $\frac{4}{3}$; the projection of this horizontal line, which is at the height $x'x_2$ above the foot ab of the plane, will be ($Prol.$ 2) parallel to ab, and at a distance from it equal to $\frac{1}{4}$ of $x'x_2$; drawing therefore xy parallel to ab, and at this distance, it will be the required projection in plan. The points x and y thus found will be two points of the projection in plan required. In the same way any number of points can be found, and the curve $axcyb$, traced through them, will be the required projection in plan.

The construction just explained, although very simple, may be abridged as follows: Through a' draw $a'f''$ perpendicular to $a'b'$; prolong $x'x'$ to the left, and set off from m''', where it cuts $a'f''$, the distance $m'''x'''$ equal to mx, as before found. Through $a'x'''$ draw the indefinite line $a'e'$. Now, to construct the projection of any other point in plan, as c on the element at the height c'; through c' draw a line parallel to $a'b'$, take the part $o'''c'''$ intercepted between $a'f''$ and $a'e'$ and set it off from o, where the projection of the element through c' cuts ab, to c along the projection of the element; c will be the required point. This is evident from the relations which the heights and horizontal distances considered bear to each other.

2d. To find the projection in plan of the intersection of the cylinders. Draw AD perpendicular to AL. If a distance Ar'' equal $x'x_2$ is set off on this line, and a parallel to AL be drawn through r'', the point u'' where it cuts the quadrant will give the point on it through which the element of the second cylinder, at the height $x'x_2$ of the two elements at x' and y', is drawn. Through u'' drawing an indefinite line parallel to AB, the bounding element of the second cylinder, it will be the projection in plan of the element at the height $Ar''=x'x_2$; and the points w and v, where it cuts the two projections of the elements of the first cylinder at the same height, will be two points of the required projection. In the same way other points would be found, by constructing the projections in plan of corresponding elements on the two cylinders.

This operation, like the former, may be also abridged as follows: Through b' draw a perpendicular $b'd'$ to $a'b'$. With a radius equal to AL describe a quadrant tangent to $b'd'$ at

b'. Now, if $x'y'$ be prolonged to the right, it is evident that the distance $r''u'''$, intercepted between $b'd'$ and the quadrant, is equal to $r''u''$. or to ru. In like manner, the point C' in plan is obtained by setting off from t, on the element AB and along oC', the distance $tC=t'C'$. The curve $AwCvB$, drawn through the points thus determined, will be the projection in plan of the intersection of the two cylinders.

3d. To make the development of the portion of the cylinder which lies between the two intersections thus determined, it will be necessary to obtain the distances of the points of these two intersections from a curve of right section: since the tangent to this curve at any point being perpendicular to the element of the cylinder at the same point, the curve when developed will also be perpendicular to the elements when developed, and will therefore develop into a right line.

To determine the relative positions of these curves in development: first develop the curve of right section $a'c'b'$, on the line $a'b'$ prolonged to the right, by setting off the distances $b'y''$, $b'c''$, &c., to a'', equal respectively to the lengths of the arcs $b'y'$, $b'c'$, &c., to a'. Through the points y'', c'', &c., draw lines perpendicular to $b'a''$; these will be the developed elements of the cylinder through y', c', &c. Set off along these lines the distances $y''v$, $c''C'$, &c., respectively equal to the distances y_2v, $o'C'$, &c., and through the points B_1, v_1, C_1, w_1, A_1, draw a curve. This is the developed intersection of the two cylinders. Make the same constructions, on the same developed elements, with respect to the distances of the points y, c, x, a, from $a'b'$; the curve bc_1a_1, drawn through these points, will be the developed intersection of the oblique inclined plane and the cylinder.

In like manner, if the given cylinder were cut by a plane perpendicular to the horizontal plane and oblique to its axis, of which ab, for example, is the trace, the developed curve of its intersection would be obtained by setting off along the developed elements the distances of the points a, m, o, &c., from $a'b'$, and through the points thus determined drawing a curve bo_1a_1.

Remark. The curve of right section in development serves only as a fixed line from which the relative positions of other points, with respect to it and to each other, can be determined; since it develops into a right line, and the elements in development are perpendicular to it. The position of this curve may be therefore fixed arbitrarily, as may be found most convenient for the purposes of the drawing.

Prob. 5, (*Pl. A, Fig.* 4). Let the semicircle $A'C'B'$ be the curve of intersection of a given semi cylinder by a vertical plane, the diameter $A'B'$ being horizontal, and suppose the bounding elements through A' and B' to be limited at a horizontal plane at the distance AA' below $A'B'$, and to pierce this plane at the points E and D, on a line parallel to AB and at a given distance from it; these elements being oblique both to the vertical plane and to the horizontal plane, and therefore not projected on either into their true lengths.

Let LM be the axis, and ED a bounding element of another semi cylinder, the right section of which is a semicircle; LM and ED being also in the given horizontal plane.

It is proposed, with these data, to find the lengths of the elements of the oblique cylinder intercepted between the vertical plane of A'B'C' *and the horizontal cylinder; the curve of right section of the oblique semi cylinder at any assumed point; and the development of the portion of it which lies between the given vertical plane and the horizontal semi cylinder.*

1st. The simplest method of finding the true lengths of the elements between the vertical plane and horizontal cylinder will be to construct their projections on another vertical plane parallel to them. Let BN be the trace of such a plane on the horizontal plane containing the axis LM of the horizontal semi cylinder, and Bc' its trace on the given vertical plane. This assumed plane cuts from the horizontal semi cylinder an ellipse of which DN is evidently the semi transverse axis, and the radius of the semi cylinder, which is equal to the distance between the axis and the bounding element ED, is the semi conjugate; setting off this distance from N to N', on a perpendicular to DN, and describing the quadrant of an ellipse DN', on these lines as semi axes, it will be the half of the curve cut from the semi cylinder, and will be its position when the plane in which it lies is revolved around its trace BN to coincide with the given horizontal plane. In this revolved position of the plane, the line Bc', in which it cuts the given vertical plane, will be found in BC'' perpendicular to BN. As this plane contains the bounding element of the oblique semi cylinder projected in BD, this element will be found in DB'' when revolved, the height BB'' being equal to BB', the height above AB in which the element pierces the given vertical plane.

Now, if any other vertical plane be passed parallel to the one assumed, as that of which $CC_{,}$, and CC', parallel respectively to BN and Bc' are the traces, it will cut from the horizontal semi cylinder an ellipse, equal to the one already described; and from the given vertical plane the line CC''; and, as it contains also the element projected in $CC_{,}$, by revolving this plane also, like the last, tracing the quadrant of the ellipse cut out, the line CC'', and the element in their revolved positions; the portion of this element between the vertical plane and the horizontal cylinder may thus be determined. In like manner, the corresponding lengths of any other elements, as those which are projected in $mm_{,}$, and $nn_{,}$, might be found. But as these successive operations would be long, a more simple and expeditious method is resorted to, as follows: As the elements of the horizontal semi cylinder are parallel to the given vertical plane, if all the points on this plane and cylinder are projected on the assumed vertical plane of which BN and Bc' are the traces, by a system of lines oblique to this plane and parallel to the elements of the horizontal cylinder, it is evident that all the ellipses cut from the horizontal cylinder will be projected into the one cut from it by the assumed vertical plane; that all the lines, as CC'', mm', nn', &c., cut from the given vertical plane, will be projected in Bc'; and, in the revolved position of the assumed vertical plane, will be found in BC''; whilst the portions of elements of the oblique semi cylinder, which lie between the horizontal semi cylinder and the given vertical plane, will be projected on the assumed vertical plane in their true lengths, and, in its revolved position, will be found parallel to $B''D$, and drawn through points n''', C'', &c., at the same height above B, on the line BC'', as the corresponding points m', C', n', are above AB. Drawing these parallels, the portions $n'''m$, $C''C''''$, &c., between the line BC'' and the curve DN', will be the lengths required.

As there are two elements on the oblique semi cylinder, one on each side of the highest one, projected in $CC_{,}$, as those projected in $mm_{,}$, and $nn_{,}$, which are of the same altitude, the lines $B''D$, $n'''m$, &c., will be respectively the revolved positions of the projections of the corresponding pairs of these elements.

Remark. By using the system of oblique projecting lines, instead of the usual mode of perpendicular ones, the relative positions of the lines projected are not changed, since these lines, being all parallel to the assumed vertical plane, will be projected on it in their true lengths, whether the projecting lines be oblique, or perpendicular to this plane. By the system of perpendicular projections, a separate construction, like the first, would have been requisite to determine each element; whereas by the oblique one used the construction of one ellipse $DN_,$ and of the line BC'' are alone sufficient.

2d. The curve of right section must lie in some plane perpendicular to the elements of the oblique cylinder. To fix such a plane, draw a line XY perpendicular to the pro jections of the elements on the horizontal plane; and from Y, where it cuts BD, another line YZ perpendicular to the projections of the elements on the assumed vertical plane. These two lines may be taken as the traces of a plane on these two planes; and, as these traces are perpendicular to the projections of the elements on the two planes, the plane itself will be perpendicular to the elements, and will cut from the cylinder a right section.

To construct this curve of right section, it will be neces- sary to find, in the first place, on the assumed vertical plane, the projections of the points in which the elements of the oblique semi cylinder pierce the plane of right section. To do this, it is evident that the vertical planes which contain these elements, as the one, for example, of which $CC_,$ and CC' are the traces, cut the plane of right section in lines parallel to YZ, its trace on the assumed vertical plane. To find the projection on this plane of the line cut out by the vertical plane of which $CC_,$ and CC'' are the traces, it is plain that the point z, where the horizontal traces XY and $CC_,$ intersect, will be one point of the required line. This point, being in the assumed horizontal plane, will be pro- jected into the line BN, the ground line of the assumed vertical plane, at z', by drawing a line through z parallel to AB, according to the method of oblique projections adopted. If from z' a line be drawn parallel to YZ, this line $z'z''$ will be the required projection of the line cut from the plane of right section by the vertical plane which contains the ele- ment projected in CC'. The point z'', where this line cuts $C''C'''$, the projection of this element on the assumed verti- cal plane, will be the projection on this plane of one point of the curve of right section. In like manner, projecting the points X, x, y, &c., into the ground line BN at X', x', y', &c., and, from these last points, drawing parallels to YZ, the points X'', x'', y'', &c., in which they cut the correspond- ing elements in projection, will give other points; and the curve $X''x''z''y''Y'$ will be the projection of the curve of right section required.

Remark. Since the elements are projected on the as- sumed vertical plane into their true lengths, it is evident that taking any point of this projection of the curve of right section, as x'' for example, the distance $x''m''$ will be the true distance of the point of which x'' is the projection from the horizontal semi cylinder, as measured on the element projected in $n'''m''$; and $x''n'''$ will be its true distance from the vertical plane containing the semicircle $A'C'B'$. In like manner, the true distances of other points of the curve of right section from this plane and from the horizontal semi cylinder measured along the elements may be found from the projection of this curve.

Having found the projection of the curve of right section, the curve itself can be found by revolving the plane of right section upon the horizontal plane around its trace XY. The distance zz_2 of the point projected in z'' from XY is evidently equal to $z'z''$, since this line is the projection, in its true length, of the one in the plane of right section drawn through the point z in the vertical plane containing the element projected in CC^1. In like manner, the distances XX_1, xx_2, yy_2, being set off from XY, along the perpendiculars to it through these points, and equal respectively to the distances $X'X''$, $x'x''$, &c., the curve $X_1x_2z_2y_2Y_1$ will be the required one. The line X_1Y_1 which corresponds to $X''Y'$ in projection will be the diameter of this curve.

3d. Having the curve of right section in its true length, as well as the elements of the oblique cylinder, and the points where they cut this curve, it will be easy to make the required developments which, for convenience, will be done on the assumed vertical plane in its revolved position. To do this set off on the line YZ, from the point Y', the distances $Y'z_3$, $Y'x_3$, $Y'X_2$ respectively equal to the arcs $Y_1y_2z_2$, Y_1x_2, &c., of the curve of right section. The right line $Y'X_2$ will be the development of this curve. Through the points z_3, x_3, &c., thus set off, draw perpendiculars to $Y'X_2$, these will be the indefinite lengths of the developed elements drawn through the points z_3, x_3, &c. From these last points set off z_3C_3, z_3C_4 respectively equal to $z''C'''$ and $z''C''''$; in like manner, set off the distances x_3n_3 and x_3m_4 respectively equal to $x''n'''$ and $x''m'''$, &c. The curves $B''C_3n_3A_1$, and $DC_4m_3E_1$, will be the developments respectively of the semicircle $A'C'B'$, in which the oblique semi cylinder cuts the given vertical plane through AB, and of the curve in which it intersects the horizontal semi cylinder. The developed portion of the oblique semi cylinder which lies between these curves and the developed positions $B''D$ and A_1E_1 of its bounding elements will be the one required.

4th. To obtain the horizontal projection of the curve in which the semi cylinders intersect. Project the points C''', m'', &c., into DN, by perpendiculars; from the foot, as $C_{,}$, of each perpendicular draw an oblique projecting line parallel to AB; the points, as $C_{,}$, $n_{,}$, $m_{,}$, &c., in which these intersect the projections of the corresponding elements, will be points in the required projection; and the curve $Dn_{,}C_{,}m_{,}E'$ will be the one required.

Theorem 1, (*Pl. A, Fig,* 3.) *If the lines* AB *and* DE, *which bisect each other and are horizontal, are the transverse axes of two semi ellipses, the planes of the two curves being vertical, and having the same semi conjugate axis, projected in the point* C, *then will the lines which join the points of the curves at the same height above the horizontal plane of the transverse axes be parallel to each other.*

Let $A'C'B'$ be the projection of the semi ellipse, having AB for its transverse axis, on a plane parallel to itself; and let the other curve be revolved about the common semi conjugate into the plane of the first and be projected into $D'C'E'$. Drawing any line, as $m'n'$, parallel to $A'B'$, it will cut the two curves at the points m', n', and o', p', at the same height above the horizontal plane; the first two being projected horizontally in m and n; the second in $o_{,}$ and $p_{,}$ in the revolved position of the second curve, and in o and p in its original position. Joining the points o and m, also p and n, then will these lines be parallel to each other, and to the lines AD and EB which join the lowest points of the two curves. For, from the properties of two ellipses having a common conjugate axis, the corresponding ordinates of the curves to this axis will be proportional to their semi transverse axes; that is,

$$cn' : cp' :: C''B' : C''E';$$
$$\text{or, } Cn : Cp :: CB : CE;$$

by substituting the equal lines in horizontal projection. But when this last proportion obtains, the lines pn and EB are parallel. In like manner, mo may be shown to be parallel to AD, and consequently to EB. The same holds true for the lines mp and on, with respect to AE and DB.

It follows from this that two semi ellipses, having the above conditions, will be the curves of intersection of two semi cylinders, the axes of which lie in the horizontal plane of the transverse axes of the curves, and are parallel respectively to the lines joining the extremities of these axes. The converse of this proposition is also evidently true, viz.: if the axes of two elliptical or circular semi cylinders lie in the same horizontal plane and intersect, and the highest eloment of each is at the same height above this plane, then will their curves of intersection be plane curves, and be projected on the horizontal plane in the two right lines which join the opposite points of intersection of the lowest elements.

Theor. 2, (*Pl. A, Fig.* 5.) *Let* $A'C'B'$ *be the vertical projection of a semi ellipse situated in a vertical plane, of which* AD, *parallel to* $A'B'$, *is the horizontal trace; and let the point* O, *on the perpendicular* OC *to* AB *which bisects it, be the horizontal projection of a vertical line through* O, *and let the semi ellipse and this vertical be taken as the directrices of a surface, generated by moving a line parallel to the horizontal plane, and in each of its successive positions touching the vertical at* O *and the semi ellipse; then will any section of this surface by a plane, as* ad, *parallel to the plane of the ellipse, be also an ellipse of which the line* ab *intercepted between* OA *and* OB, *will be one axis, and the line* O'C' *equal to the other semi axis. And a tangent line drawn to this ellipse at any point, as the one projected in* n, n', *will pierce the horizontal plane in a line* OD, *drawn from the point* O *to the point* D, *where a tangent to the directing ellipse at the point projected in* m, m', *at the same height as* n', *also pierces the horizontal plane.*

1st. Through the point projected in *n, n'*, draw the projections *Om* and *o'm'* of an element of the surface, and project the line *ab* into *a'b'*. As the lines *BC* and *bo* are parallel, there obtains

$$BC : bo :: mC : no;$$

but $BC = B'O'$; $bo = b'O'$, and $mC = m'o'$; $no = n'o'$;

therefore, $B'O' : b'O' :: m'o' : n'o'$,

which shows, from the properties of ellipses having a common axis, that the curve $b'n'C'$ is an ellipse.

2d. Let tangents be drawn to the two ellipses at the corresponding points *m'* and *n'*, at the same height above $A'B'$. These tangents will intersect the common axis at the same point E', and the other axis at points D' and d' such, that

$$O'D' : O'd' :: m'o' : n'o'.$$

Projecting the points D' and d' into the respective planes of the two ellipses at D and d, and observing that $O'D' = CD$; $O'd' = od$; $m'o' = mC$; and $n'o' = no$; there obtains

$$CD : Cm :: od : on;$$

that is, the line joining the points D and d passes through the point O.

Remark. This surface is a *right conoid* of which the horizontal plane is the *plane director*.

Prob. 6. *To construct a tangent plane and a normal line to the conoid at any point, as* n, n′.

Draw the projections of the element *Om*, *o′m′* of the surface at the given point. Find where the tangent to the directing ellipse at the point *m*, *m′* pierces the horizontal plane. Join this point *D* with *O*. Through *n* draw a parallel to *CD*, and where it cuts *OD*, at *d*, draw a line *XY* parallel to *Om*. This is the horizontal trace of the tangent plane at *n*, *n′*. For the tangent line at the given point to the ellipse projected in *ab* pierces the horizontal plane at *d*, this is therefore one point of the horizontal trace required, and as the element of the surface is contained in the tangent plane and is horizontal the trace *XY* will be parallel to *Om*.

The line *zv*, drawn through *n* perpendicular to *XY*, is the indefinite projection of the normal to the surface at *n*, *n′*.

Prob. 7, (*Pl. A, Fig.* 6.) *To draw a tangent plane and a normal line to a helicoidal surface at a given point on the surface.*

Suppose *XYZ*, the involute of any given curve *xyz*, to be the base of a vertical cylinder; and let the line *Zy′*, tangent to this curve at *Z*, be the horizontal trace of a plane tangent to the cylinder along the element projected in *Z*; and in this tangent plane, revolved on the horizontal plane, let any inclined line *Zy″* be drawn through the point *Z*. If the tangent plane be now returned to its vertical position, and be wrapped around the vertical cylinder, the inclined line *Zy″* will form a helix on the cylinder, and the points *y″*, *m″*, &c., in their position on the cylinder, will be projected into its base, at the points *Y*, *m*, &c.; such that the arcs *ZY*, *Zm*, will be equal to the distances *Zy′*, *Zm′*, &c., from *Z* to the projection of these points in the tangent plane into its horizontal trace. If a right line be moved along this helix so that in all its positions it shall be parallel to the horizontal plane, and be projected on this plane in lines as *Zz₁*, *Yy*, normal to the curve *XYZ*, this line will generate a helicoidal surface, the elements of which will be normal to the cylinder of which *XYZ* is the base, and tangent to the one of which *xyz₁* is the base.

Let the point projected in *Y*, and which is at the height *y′y″* above the horizontal plane, be the one at which it is required to construct a tangent plane, and a normal line to the surface. As the helix makes a constant angle, equal to the one *Zy″y′*, with the elements of the cylinder, at the points where they intersect; and as a tangent to the helix at any point makes the same angle as the helix does at this point with the element of the cylinder; it is evident that the tangent to the helix, at the point projected in Y, will pierce the horizontal plane, at a point z, on the tangent drawn to the curve at Y, at the same distance from this point as the distance $Zy' = ZY$. This point z will therefore be one point of the horizontal trace of the required tangent plane. But, as the tangent plane contains the horizontal element of the surface at the given point, its trace on the horizontal plane will be parallel to this element. Drawing a line AB through z parallel to Yy, this will be the trace of the required plane. The normal will evidently be projected in Yz, as this line is perpendicular to AB the required trace.

To find the true length of this normal let CD, parallel to Yz, be the trace of a vertical plane. The element through the given point will pierce this plane at a height $Y'Y''$ above CD equal to $y'y''$. The normal line will be projected on this plane in its true length, and will pass through the point Y. But, as the vertical plane through CD is also perpendicular to the tangent plane at the given point, the line Yz' will be the projection as well as the trace of the tangent plane on this vertical plane. The line YS drawn perpendicular to Yz' will therefore be the indefinite projection of the required normal.

Remark. The traces of all tangent planes to the surface, at points on the element considered, will evidently be parallel to Yy, the projection of this element; and the trace of any one may be readily found by means of the point z. For, through y, where the projection of the element is tangent to the curve xyz, through the point z draw an indefinite line $yy_{,}$; then if at any point on the projection of the element Yy a line be drawn parallel to Yz, the projection of the tangent at Y, the point where this parallel cuts yy, will be a point in the horizontal trace of the tangent plane at the assumed point. This may be readily proved as follows: Let the tangent projected in Yz, and the element of the cylinder projected in y be taken as the directrices of a warped surface of which the horizontal plane is the plane director This warped surface will be tangent to the helicoidal surface throughout the entire element projected in Yy; for they have the same plane director; a common tangent plane at the point projected in Y; and also one, which is vertical, at the point projected in y. Now, as the warped surface in question is a hyperbolic paraboloid, any vertical plane parallel to Yz will intersect it in a right line which will cut the element projected in Yy, and pierce the horizontal plane in the line yy, where the trace of the assumed vertical plane cuts it. This point will therefore be a point in the trace of a tangent plane to both surfaces at the point where the line cut from the hyperbolic paraboloid intersects the element common to the two surfaces, and along which they are tangent.

Prob. 7, *Pl.* 1. *To construct the projections and true dimensions of the bounding lines and surfaces of the vous soirs of a horizontal full centre arch.*

This problem comprises several cases, according to the character and the positions of the surfaces which form the ends or heads of the arch.

Case 1st. *The arch being terminated at one end by a vertical plane oblique to its axis and at the other by a vertical plane perpendicular to the axis.*

Let the semicircle (*Fig.* 1) of which $B'C'$ is the diameter, be the right section of the soffit of the arch, and having divided it into any odd number of equal parts, as five for example, draw through the points of division E', F', &c., radii which prolong to the semicircle described on $A'D$ as a diameter. These radii will be directions of the joints in the right section, $A'B'$ being their common thickness. Through the points A', I', G', &c., drawing vertical and horizontal lines, they will be the exterior bounding lines of the voussoirs in right section; the line $K'K'''$ which bounds the top of the keystone being assumed at pleasure.

Let AD (*Fig.* 2) be the trace on the horizontal plane containing the axis and lowest elements of the arch of the vertical plane oblique to the axis which forms the front end of the arch, and ad the trace of the one perpendicular to the axis that bounds the back end.

As the edges of any voussoir, as the one of which $Q'M'$ $N'O'P'$ (*Fig.* 1) is the right section, are all parallel and horizontal, they will be projected into their true lengths in plan between the two traces of the end planes. The one corresponding to M' will be projected in plan (*Fig.* 2) in $M_{,}m_{,}$; that corresponding to N' in $N_{,}n_{,}$, &c.

As all the joints, except the two lowest, are oblique to the horizontal plane, their horizontal edges alone are projected into their true dimensions in plan. The two lowest, corresponding to $A'B'$ and $C'D$, are projected into $AabB$, and $CcdD$ respectively. To find the true dimensions of the others, and the development of the soffit, develop in the first place the right section of the soffit, by setting off the distances $B'E'$, $E'F'$, &c., (*Fig.* 5) respectively equal to the arcs $B'E'$, &c. (*Fig.* 1); and then, by *Fig.* 2, *Pl.* $A_{,}$

find the points B, $E_{,}$, $M_{,}$, &c., of the developed curve in which the soffit is intersected by the oblique plane of the front end; also the right line a, $e_{,}$, $m_{,}$, &c., which is the development of the right section in which it is cut by the plane of the back. The surface bounded by the developed curves and the two lowest elements of the semi cylinder is the developed soffit.

To find any oblique joint in its true dimensions, as the one projected in plan between the lines $M_{,}m_{,}$ and $N_{,}n_{,}$, set off from M', (*Fig.* 5) to the right along $M'C'$ a distance equal to $M'N'$ (*Fig.* 1), the breadth of the joint. From the point N' (*Fig.* 5) draw a line perpendicular to $B'C'$, and from N' set off $N'N_{,}$ equal to $n'N_{,}$(*Fig.* 2). Join the points $M_{,}N_{,}$; the trapezoidal figure $M_{,}N_{,}n_{,}m_{,}$ is the required joint. In like manner the true dimensions of all the other joints are found as shown on *Fig.* 5.

Case 2d. *The arch being terminated at one end by a plane oblique to its axis and to the horizontal plane of its two lowest elements and at the other by a vertical plane perpendicular to the axis.*

Let AD (*Fig.* 2) be the horizontal trace of the oblique plane, and let the angle which it makes with the horizontal plane be ?. Drawing any line rw parallel to AD (*Prob.* 1) as the projection in plan of a horizontal line of this plane, and from any point as A a perpendicular Ax to rw, the height of the point x above the horizontal plane will be three times the distance Ax. Setting off from A' (*Fig.* 3) the length Ax to x', at x' erecting the perpendicular $x'x''=3A'x'$ and joining A', x'', the angle $x'A'x''$ will be the angle between the oblique and horizontal planes. Taking now the distance Ay (*Fig.* 2) in which the line vw cuts the line Aa, parallel to the axis, and setting it off from A' to y' (*Fig.* 3), erecting a perpendicular $y'y''=x'x''$, and joining $A'y''$, the angle $y'A'y''$ will be the one between the oblique and horizontal planes measured in the vertical plane parallel to the axis of which Aa (*Fig.* 2) is the trace.

Having drawn the line $A'y''$, and the vertical at A', the projections in plan of the points in which the oblique terminating plane cuts the bounding horizontal lines of the voussoirs corresponding to the points E', I', F', &c. (*Fig.* 1) in right section are readily determined by applying the constructions in *Prob.* 4. The points in plan corresponding to E' and I' (*Fig.* 1), for example, will be found by drawing lines through E', I' parallel to $A'y'$, taking the lengths $E''E'''$, $A''A'''$ and setting them off from $E_{,}$ and $I_{,}$ (*Fig.* 2) from the trace AD to E and $I_{,}$ along the projections in plan of the corresponding edges of the voussoirs. The curve *BEF*, &c., thus obtained, and the pentagonal figures *EHGF*, &c., are the projections in plan of the oblique sections of the soffit and voussoirs of the arch.

Remark. As a verification of the accuracy of the constructions, the lines *IE*, *GF*, &c., prolonged should pass through the point *L* where the axis cuts the trace *AD;* and the lines *A,I*, *HG*, &c., should be parallel to *AD*, as the top surfaces of the voussoirs are horizontal planes.

Case 3d. *The arch being terminated by an oblique plane, as in either of the preceding cases, and at the other extremity by a semi circular cylinder, its axis and two bounding elements being in the same plane as the corresponding lines of the arch and perpendicular to them.*

Let *ad* (*Fig.* 2) be one bounding element of the given semi cylinder. Having set off the radius of this cylinder from *D*, (*Fig.* 4) on the line *A'D* prolonged, and described a portion of the semicircle tangent to the vertical *DN''*, draw through the points *M'*, *N'*, &c., (*Fig.* 1) parallel lines to *A'D*. By *Prob.* 4 find the projection in plan of *bef*, &c., of the curve of intersection of the soffit of the arch and the semi cylinder; also the projections of the points *i*, *h*, *g*, &c., in which the horizontal edges of the voussoirs intersect the semi cylinder; the pentagonal figures *efghi*, &c., will be the projections in plan of the lines in which the surfaces of the voussoirs intersect the semi cylinder. Any point in plan, as *n*, is found by setting off a length *n,n* from *ad* along the projection of the edge corresponding to *N'*, equal to the distance *N''N'''* (*Fig.* 4).

Remark. The lines *ei*, *fg*, &c., are portions of ellipses, which prolonged pass through the point *l*, in which the axis of the arch cuts the bounding element *ad*. The lines *gh*, *no*, &c., are right lines, being the projections of the intersection of the horizontal surfaces of the voussoirs with the semi cylinder. The lines *hi*, *op*, &c., are the projections of arcs of circles in which the side vertical planes of the voussoirs corresponding to *H'I'*, *O'P'*, intersect the semi cylinder.

The true dimensions of the joints in either of the two last cases are found by setting off from the line *B'C'* (*Fig.* 5) the lengths along the perpendiculars, at the points *E'*, *I'*, &c., which correspond to the distances respectively of the points *E*, *I* and *e*, *i*, (*Case 3d*, *Fig.* 2) from *A'D*.

Remark. As a verification of and aid to accuracy of construction, let a line *Ll* (*Fig.* 5) be drawn parallel to the edge *M'm* and at a distance from it equal to *L'M'* (*Fig.* 1) the radius of the right section; the right lines MN and $M_{,}N_{,}$ prolonged should intersect the line Ll at the same point L, such that the length Ll (*Fig.* 5) shall be the same as Ll (*Fig.* 2). In like manner the curve $nu_{,}m$ prolonged should intersect the same line Ll at the point l which corresponds to the one l (*Fig.* 2). Similar constructions of verification will be found on *Fig.* 5 (*Pls.* 3 and 4).

From an examination of *Fig.* 2 it will be seen, that the projection in plan of the voussoir corresponding to $M'N'O'$, &c., (*Fig.* 1) in right section, will in *Case* 3d be bounded on one end by the figure MNO, &c., on the other by the one mno, &c.; and by the parallel lines which join the corresponding points Mm, Nn, &c.

Application. Having found the principal dimensions of the voussoirs, let it be required to cut from a single block of stone of the form of a rectangular parallelopipedon the voussoir corresponding to $M'N'O'$, &c.. (*Fig.* 1) in *Case* 3d. Drawing a line $S'T'$ through Q' perpendicular to $O'P'$, and prolonging $O'N'$ to R' on the line drawn through M' parallel to $O'P'$, the rectangle $R'T'$ will evidently be the dimensions of the end of the block within which the voussoir in right section can be inscribed. The dimensions of the length of the block will evidently be determined by drawing through the point Q (*Fig.* 2) a line $R'Q$ parallel to ro. Having inscribed on the end of a block of the form and dimensions thus found, the figure in right section, the block would be prepared by cutting away those portions, as $M'S'Q'$, &c., which are exterior to the figure. This being done, the points corresponding to M, N, O, &c., and m, n, o, &c., (*Fig.* 2) can be set off on the corresponding edges, and the two ends of the voussoir, the one terminated by the oblique plane, the other by the semi cylinder, be obtained. In cutting away the portions of the block to form the curved surfaces of the soffit and of the end of the voussoir, a model cut from a thin board, by shaping it on the back to the form of the arc $M'Q'$ (*Fig.* 1), and a like model cut to the form of the arc DN'''', would be requisite as a guide to the workman, to be applied, from time to time, in a direction perpendicular to the elements of the cylinders, until it is found that the models coincide accurately at all points with the prepared surfaces.

Models also of the true forms of the joints, determined in *Fig.* 5, may be cut from thin pasteboard, or any like material, and be used to verify the work. These last would evidently not be requisite to guide the workman in setting off his points where he works from a block of the above form. But, in cases where a block of irregular shape has to be taken, they may be found the most convenient for setting off the points to determine the form of these joints on the stone.

Prob. 8, *Pl.* 2. *To construct the projections and true dimensions of the voussoirs in the groined and cloistered arches.*

In each of these cases the soffits of the arches are formed by the intersections of two semi cylinders, the axes of which are in the same horizontal plane, and their top elements at the same height above this plane. From these conditions, the curves of intersection of the soffits (*Theor.* 1) will be plane curves, and will be projected in plan in the diagonal lines which join the intersections of the lowest elements of the semi cylinders.

In the cases selected to illustrate this problem (*Pl.* 2), the curve of right section of one of the semi cylinders is a semicircle (*Fig.* 1), that of the other (*Fig.* 2) a semi ellipse, each having the same rise $L'K'$; and their axes are taken perpendicular to each other. The joints in each arch corresponding to $F'G'$, &c., are normal to the soffit, or surfaces of their respective cylinders; the upper and lower edges of the corresponding joints in each arch being in the same horizontal plane, as well as the top surfaces, as $G'H'$ (*Figs.* 1, 2), of the voussoirs.

Remark. *Fig.* 1 is the right section of the semi circular arch; *Fig.* 2 that of the semi elliptical arch; *Fig.* 3 above the line *ab* is a portion of the plan of the groined arch, the soffit of the semi circular arch being projected within the angle BKC and the corresponding soffit of the semi elliptical arch, only the half of which is shown in plan, being projected within the angles aKB and bKC. *Fig.* 5 on the right represents two of the joints of a groin stone belonging to the semi circular arch with the development of the portion of the soffit between them; that on the left the true dimensions of the corresponding parts of the same stone which forms a part of the other arch.

Fig. 4 below the line *ab* is a portion of the plan of the cloistered arch, the soffit of the semi elliptical portion being projected within the angles $B_{,}KC_{,}$; that of the semi circular portion being projected within the angles $B_2KC_{,}$ and $B_{,}KC_2$. *Fig.* 6 on the right represents the two joints of a groin stone which forms a part of the semi circular arch with the portion of the soffit between them; that on the left the corresponding parts of the same stone of the other arch.

Groined Arch. Having constructed (*Fig.* 1) the right section of the semi circular arch as in *Prob.* 7, assume $B'C''$ (*Fig.* 2) as the transverse axis of the ellipse of right section of the other arch, and placing it at any convenient position, on the left, perpendicular to the direction $B'C''$ (*Fig.* 1), set off the semi conjugate $L'K'$ equal to the radius of the semicircle, and describe the semi ellipse. Find on the semi ellipse, as shown by the lines $X''h'$, &c., on the left of *Fig.* 1, the points E', F'', &c., at the same height above $B'C''$, as the corresponding points of the semicircle. Construct tangents to the semi ellipse at these points, and at the same normals for the directions of the joints. Find on these normals the points I', G', &c., at the same heights as the like points in *Fig.* 1. Through the points I', G', (*Fig.* 2) draw the vertical and horizontal lines $I'II'$ and $G'II'$, for the bounding lines of the voussoirs. Having completed, in this way, the right section (*Fig.* 2), draw, in plan (*Fig.* 3), the projection of the axes, and the bottom elements of the arches, corresponding to the points B', C'' (*Figs.* 1, 2). Drawing the diagonal lines BB' and CC' between the points where these elements intersect in plan, they will be (*Theor.* 1) the projections of the ellipses in which the two semi cylinders intersect, and which form the edges of the groins.

To find the projections in plan of the voussoir of the groin which corresponds to the one $M'N'O'$, &c. (*Fig.* 1), and $F'G'II'$, &c. (*Fig.* 2), draw Mm (*Fig.* 3), the projection of the lower edge of the joint corresponding to M'' (*Fig.* 1), and Mm, the corresponding projection for the point F' (*Fig.* 2); and in like manner the lines Nn and Nn, the projections of the upper edges. Joining M and N gives the projection of the intersection of the planes of the two joints. Find in like manner Qq and Qq; Pp and Pp, the projections of the edges corresponding to the joints $Q'P'$ and $E'I'$. Joining Q and P gives the projection of the intersection of the planes of the two last joints. Having found the projections in plan of the edges of the joints and their intersections, the voussoir is terminated on the semi circular arch by a joint of right section mp, taken at any suitable distance from the point P'; and on the semi elliptical arch by a like joint $m'p'$. The required voussoir in plan will be the figure $MmpP'p'm'$. The part above the line MC' belonging to the semi circular arch; that to the right of it to the other.

To find the true dimensions of the joints of this voussoir and of the portion of the soffit which belongs to the semi circular arch, draw a line mp (*Fig.* 5) to correspond to the one

mp (*Fig.* 3) of the plane of right section by which the vous soir is terminated. On this line set off $mn = M'N'$ (*Fig.* 1) the breadth of the upper joint; $nq = M'Q'$ the length of the arc between the joints; and $qp = Q'P'$ the breadth of the lower joint. Through these points draw perpendiculars to the line, and set off on them $nN = mM$ (*Fig.* 3); $mM = nN$; $qQ = qQ$, and $pP = pP$. Join the points M, N, and Q, P, by right lines; the points N, Q, by a curve line, an intermediate point of which can be found by constructing in plan the element yx of the soffit which corresponds to the middle point x' (*Fig.* 1) of the arc $M'Q'$ and setting this line off on *Fig.* 5 from y, the middle point of nq to x. The *Fig.* 5 will give the joints required in their true dimensions, also the developed portion of the soffit between them, and which is bounded at one end by the curve of right section corresponding to nq, and at the other by the portion of the ellipse of the groin corresponding to MQ (*Fig.* 3).

In like manner the true dimensions of the joints, and the portion of the soffit between them, which belong to the same stone, and form the portion of the elliptical arch terminated by the joint of right section $m,p,$ (*Fig.* 3) may be found, as shown (*Fig.* 5) on the left; by setting off the distances $p,q,$, $q,m,$, $m,n,$, respectively equal to $E'I'$, $E'F'$ and $F'G'$(*Fig.* 2). and on the perpendiculars to $p,n,$, through the points $p,$, $q,$, &c., setting off the distances p,P', q,Q, &c., respectively equal to p,P &c., in plan (*Fig.* 3).

The portion of the keystone which forms the top of the groins at the point X' (*Fig.* 3), is limited on the semi circular arch by the joint of right section $G,N,$; and by the one Hh on the semi elliptical cylinder, with a corresponding one on the right of K.

The joints of right section of the different courses, as mp and $G,N,$, are arranged to break joint.

Cloistered Arch. The constructions for determining the projections, &c., of the joints and their true dimensions, are precisely the same in this case as in the preceding.

In *Fig.* 4 aa,b,b are the exterior lines in plan, and B,C,B,C_2 the interior lines of the top of the walls, or the imposts of the arches; the semi circular arch springing from the lines $B_2C,$ and $C_2B,$ and the semi elliptical form $CB,$. The lines C,K and B,K are the projections of the half of each groin.

By drawing in plan the edges of the joints corresponding to $P'Q'$ and $M'N'$ (*Figs.* 1, 2), and joining the points P, Q, and M, N (*Fig.* 4), the intersection of these joints are obtained in plan. The groin stone which corresponds to these joints is limited by a plane of right section *mp* (*Fig.* 4) taken at pleasure on the semi circular arch, and a like one *m,p,* on the semi elliptical. All the parts of this groin stone will therefore be projected within the figure *mpPp,m,M.*

To construct the joints and the portion of the soffit in their true dimensions which belong to this stone, commence (*Fig.* 6) by setting off on a right line the distances *nm, mq,* and *qp,* respectively equal to *N'M', M'Q',* and *Q'P'* (*Fig.* 1), and which correspond to the joint of right section *mp* (*Fig.* 4). Draw through the points *n, m,* &c., thus set off, perpendiculars to *np,* and along these perpendiculars set off the distances *nN, mM, qQ,* and *pP,* respectively equal to the same lines on *Fig.* 4. Join the points *NM* and *QP* by right lines; and *MQ* by a curved line, an intermediate point of which corresponding to *x'* (*Fig.* 1) is found by setting off from *y,* the middle of *mq* the distance *y,x,* equal *y,x,* (*Fig.* 4).

In the same way the corresponding portions of the groin stone belonging to the joint of right section *m,p,* on the semi elliptical arch are found from *Figs.* 2 and 4.

The top groin stone at *K* (*Fig.* 4) which forms a portion of the two arches is represented as a single stone.

The joints of right section in the different courses of voussoirs are arranged as shown in plan to break joint.

Application. Having determined the projections in plan of the edges of the joints of a groin stone with the true dimensions of the joints, and the portion of the soffit of each arch belonging to it, their uses in shaping the stone from the solid block will be easily understood. Taking, for example, the groin stone of the groined arch, the right sections of which are given in *Figs.* 1 and 2; the plan in *Fig.* 3; and the true dimensions of the joints, &c., in *Fig.* 5, it will be readily seen that, supposing a block from which the stone is to be shaped to be a rectangular parallelopid, its thickness must be such that the right section *M'N'O',* &c., (*Fig.* 1) can be inscribed within the rectangle of the end that corresponds to the joint of right section *mp* (*Fig.* 3) of the semi circular arch, and the figure *F'G'H',* &c., be inscribed within the end corresponding to the joint *m,p,* of the semi elliptical arch; and the length of the block must be equal to *m,M,* and its breadth to *Mm,.*

·Having inscribed upon the ends of the block the two figures of cross section, the portions of the solid exterior to them are gradually worked off until the dressed surfaces coincide with *Fig.* 5, which may be ascertained either by measurement, or by the actual application of these figures cut from some thin flexible material.

Remark. A careful examination of the lines of the figures will show the geometrical methods for determining the tangents to the points on *Fig.* 2 which correspond to *Fig.* 1.

Prob. 9, *Pls.* 3, 4. *To construct the projections and true dimensions of the voussoirs of the rampant cylindrical arch.*

This problem, which comprises two cases, is a variation of *Prob.* 7. The arch in this, as in *Prob.* 7, being terminated at one end by a vertical plane, and at the other by a horizontal semi cylinder having its axis and elements parallel to the vertical plane of the end; the elements of the soffit and the edges of the voussoirs of the arch being oblique both to the vertical plane and to the horizontal plane of the plan.

Case 1, *Pl.* 3. Construct (*Fig.* 1) the semicircle $B'T'C'$, to represent the oblique section of the arch by the vertical plane of the end. Let ad (*Fig.* 2), assumed at any convenient distance from $A'D'$, be the projection in plan of the lowest element of the semi cylinder of the other end, which with the axis is taken in a horizontal plane at the distance $a'''a''$ (*Fig.* 3) below the line $A'D'$; and let the lowest elements of the arch drawn from the points $A'B'C'D'$, and its axis from L', (*Fig.* 1) be taken to intersect the line ad, and to lie in vertical planes perpendicular both to the vertical plane of the end and to the horizontal plane. The edges of the voussoirs as $A'a$, $B'b$, &c., in plan will be perpendicular to $A'D'$.

To obtain the edges of the voussoirs in their true dimensions, it will be necessary to find their projections, as in *Prob.* 5, on a plane parallel to them.

Let the vertical plane which contains the edge projected in $A'a$ be taken for this purpose, and suppose it revolved around the line $A'K'''$, its trace on the vertical plane of the end, to coincide with this plane. In this new position of the side vertical plane the edge projected in $A'a$ will be obtained on it by setting off $A'a'''=A'a$; erecting at a''' a perpendicular $= a'''a''$, and joining $A'a'''$. The edges drawn from B', C', D', and the axis of the arch will all evidently be projected into the same line $A'a''$.

The projections of the other edges on this plane will evidently be parallel to $A'a''$ (*Fig.* 3), and their positions will be found by drawing through the points E', I', &c., (*Fig.* 1) lines parallel to $A'D'$, and from the points E'', &c., in which they cut $A'K'''$ drawing parallels to $A'a''$ (*Fig.* 3).

As a'' is the revolved position of the point in which the vertical side plane cuts the lowest element of the semi cylinder of the end, and as the axis is in the same horizontal plane as this element, by setting off on the horizontal through a'', and to the left of a'', the radius of the semi cylinder and from the point thus set off describing an arc $a''X'$, it will be the one cut from the semi cylinder by the vertical side plane. The lines, as $H''H'''$, $E''E'''$ intercepted between the arc $a''X'$ and the line $A'K'''$, are the projections in their true dimensions of the required edges of the voussoirs.

To obtain a right section of the arch, for the purpose of constructing the joints of the voussoirs in their true dimensions, and the development of the soffit of the arch; from the point A' draw the line $A'Y'$ perpendicular to the projections of the edges on the vertical side plane; this line may be regarded as the trace of a plane of right section on the side vertical plane, and the line $A'D'$ as its trace on the horizontal plane through $A'D'$. If the plane of right section thus fixed be revolved around $A'D'$ until it becomes horizontal, the right section contained in it will be determined in its true dimensions. The points as e', f', &c., (*Fig.* 4) will be found, after this revolution, by setting off from the line $A'D'$ along the projections of the elements in plan corresponding to the points E'', F'', &c., distances equal to $A'e'$, $A'f'$, &c., (*Fig.* 3) measured from the point A' along the line $A'Y'$. The curve $B'e'f'm'q'C'$ (*Fig.* 4) thus determined is the curve of right section of the soffit, and the figure $m'n'o'p'q'$ the right section of the voussoir corresponding to $M'N'O'P'Q'$ (*Fig.* 1) of the end.

Having obtained the right section, the development of the soffit and the joints in their true dimensions are found, as in *Probs.* 5 and 7, as follows: Having drawn a line $B'C'$ (*Fig.* 5) set off along it the distances $B'e'$, $e'f''$, $f'm'$, &c., respectively equal to the arcs $B'e'$, $e'f'$ &c., on the curve of right section (*Fig.* 4). The right line $B'C'$ will be the development of the curve. Through the points B', e', f', &c., draw perpendiculars to $B'C'$ which will be the elements of the soffit in development, which are the lower edges of the joints and correspond to the points B', E', &c., (*Fig.* 1). As the true distances of the extremities of these edges from the plane of right section are given on *Fig.* 3, and are the distances $e'e''$, $e'E'''$ for the edge projected in its true length between the line $A'K'''$ and the curve $a''X'$; by setting off $e'e''$ and $e'E'''$, from e' to E'', and e' to E''' (*Fig.* 5), the points E'' and E''' will be respectively points of the development of the curves in which the soffit intersects the vertical plane of the one end of the arch and the horizontal cylinder that terminates the other. In like manner the points F', M', &c., of the developed curve $B'T'C'$ are found, and those as b'', E''', &c., of the other end. To obtain the true dimensions of any joint, as the one corresponding to $E'I'$ (*Fig.* 1), set off the distance $e'i'$, (*Fig.* 5) equal to the breadth of the joint in right section, which is $e'i'$ (*Fig.* 4). Through i' draw a perpendicular to $B'C'$, and set off from it the distance $i'I'$, $i'A'''$, respectively equal to $a'A''$, $a'A'''$ (*Fig.* 3). Join I', E', by a right line, and $A'''E''''$ by a curved line, the figure $A'''I'E'E'''$ will be the required joint in true dimensions. The other joints are found in like manner.

To find the dimensions of a block of the form of a rectangular parallelopiped from which one of the voussoirs, as the one corresponding to $M'N'O'P'Q'$ could be cut, it will be observed that the edges of this voussoir are projected in their true dimensions on *Fig.* 3, between the line $H''H'''$, which corresponds to the points N', O' (*Fig.* 1), and the line $E''E'''$, which corresponds to Q'; drawing therefore from H'' a perpendicular to $E''E'''$ prolonged, and from E'' one to $H''H'''$ prolonged, the rectangle thus formed will be the true dimensions of one side of the block. As $m'n'o'p'q'$ (*Fig.* 4) is the cross section of the same voussoir, the breadth of the rectangle of the end of the block must be equal to $m'p'$, in order that the figure $m'n'o'$, &c., can be inscribed within it. The manner of setting off the different lines on the block, with the view of dressing it into shape, will be readily seen from what has already been stated on this point in the preceding problems.

Remarks. From the preceding constructions, the joints and the development of the soffit for any other plane end passing through the line $A'D'$, and having the line $A'Z'$, for example, for its trace on the vertical side plane, can be readily found, by setting off, on *Fig.* 5, from the line $B'C'$, the distances between the corresponding points on the lines $A'Y'$ and $A'Z'$, as $h'h''$, for example, in the same way as for those between $A'Y'$ and $A'K''$, and through the points e'', f', &c., drawing the developed curve of intersection of the soffit and assumed plane, and constructing the corresponding joints as $e''i''A''E''$.

Prob. 9, *Case* 2*d*, *Pl.* 4. This case is a variation of the preceding one, the axis and elements of the soffit of the arch being oblique both to the horizontal plane, and to the vertical plane which terminates the arch at one end, but situated in vertical planes oblique to the vertical plane of the end. The position of the semi cylinder which terminates the other end is the same in all respects as in the preceding case.

Fig. 1 represents the end of the arch in the vertical plane. The curve of the soffit $B'F'M'C'$ in this plane is a semicircle. The arch for the mere illustration of this case consists of only three voussoirs.

Fig. 2 represents the projections of the elements in plan; the axis of the arch and the lowest elements of its soffit intersect the lowest element of the semi cylinder, which is projected in the line ad, parallel to $A'D'$, and lies in a horizontal plane at the distance $D'D''$ below $A'D'$, at the points b, l and c.

Fig. 3 represents the projections of the edges of the voussoirs on the vertical side plane, parallel to them, of which $D'T$ and $D'V'$ are the traces on the horizontal plane of the plan, and the vertical plane of the end. The system of projecting lines in this case is the same as the one used in *Prob.* 5.

Fig. 4 represents the revolved position, on the horizontal plane, of the right section of the arch, contained in the plane of right section of which XY', perpendicular to the projection of the axis $L'l$ is the horizontal trace, and Yz', perpendicular to $D''d$, the projection of the axis on the side vertical plane, is the trace on this plane.

Fig. 5 represents the joints and the development of the soffit in their true dimensions.

Having constructed *Figs.* 1 and 2, find, by *Prob.* 5, the projections of the edges of the joints on the vertical side plane of which $D'T$ is the horizontal trace, assuming, in the first place, the line $D''d$, as the projection of the axis and lowest elements on this plane, and parallel to which all the other projections of the edges are drawn; the one corresponding to the point M' (*Fig.* 1), for example, is found by setting off from D'' (*Fig.* 3) on the revolved position $D'K'''$ of the trace of the side vertical plane with that of the end, the distance $D''M''$ equal to MM' (*Fig.* 1), and drawing $M''M'''$ parallel to $D''d$.

Representing, by the line drawn through T parallel to ad, the axis of the semi cylinder, the ellipse cut from the semi cylinder of the end will have dT, and TS' equal to the radius of the semi cylinder, for its semi axes. Having described the quadrant dS' of the ellipse in its revolved position, the projections of the edges of the joints intercepted between it and the line $D'K'''$ will be the true lengths of the edges.

To obtain the right section, take XY perpendicular to the axis $L'l$ in plan, and Yz' perpendicular to its projection $D''d$ on the vertical side plane, as the traces of the plane of right section. Having found (*Fig.* 4, *Pl. A,*) the projections c', $n_{,}$, n', &c. (*Fig.* 3) of the points in which the elements of the soffit and the edges of the joints pierce the plane of right section, next construct from these (*Fig.* 4) the right section as revolved on the horizontal plane.

To construct the soffit and joints in their true dimensions, draw (*Fig.* 5) a line $a_{,}D'$ (*Fig.* 4, *Pl. A*) and set off on this from $b_{,}$ to $c_{,}$ the length of the curve of right section $b_{,}c_{,}$ (*Fig.* 4). Drawing through the points $b_{,}, f_{,}, m_{,}, c_{,}$, perpendiculars to $a_{,}D'$, set off on them above and below $a_{,}D'$ distances $m_{,}M'$, $m_{,}M'''$, respectively equal to $m_{,}M''$ and $m_{,}M'''$ (*Fig.* 3). The curves $B'F'M'C'$, and $bM'''c$, drawn through the points thus obtained, will be the developments of the intersection of the soffit with the end plane and semi cylinder. To construct the joint of which $M'M'''$ is one edge, set off $m_{,}n_{,}$ on $a_{,}D'$ equal to $m_{,}n_{,}$ (*Fig.* 4), the width of the joint in cross section; through $n_{,}$ draw a parallel to $M'M'''$, and set off on it $n_{,}N'$, $n_{,}N'''$, respectively equal to $n'N''$ and $n'N'''$ (*Fig.* 3). Joining $M'N'$ by a right line, and $M'''N'''$ by a curve line, the figure obtained is the required joint. The others are found in like manner.

Remark. A comparison of the lines on *Fig.* 5 with those on *Fig.* 3 will point out the manner of constructing an intermediate point as v of the curve $M'''N'''$.

Prob. 10, *Pl.* 5. *To construct the projections and true dimensions of the voussoirs of the hemispherical dome.*

Let the semicircle (*Fig.* 1) $B'L''C'$ be the vertical section of the soffit of the dome, and suppose it divided into seven equal parts at the points E', F', &c. Drawing radii through the points E', &c., set off upon them the equal distances $E'I'$, $F'G'$, &c., and complete, as in the preceding cases, the figures $E'I'H''G'F'$, &c., to represent the sections of the voussoirs. If *Fig.* 1 be supposed to be revolved about the vertical radius $L'L''$ as an axis, any section of a voussoir, as $E'I'H''G'F'$ would generate the entire voussoir of the dome comprised between the horizontal circles on the soffit projected in the lines $E'Q'$ and $F''M'$. The lines $E'I'$, $F'G'$, in this revolution generate the joints between the voussoir in question and the two in contact with it, which joints are portions of a cone of which the centre of the dome is the vertex, and $L'L''$ the axis. The line $I'H'$ will generate a cylindrical surface having the same axis, and the line $H'G'$ a plane.

Having in this manner determined the bounding surfaces of each course of voussoirs, the course is divided into blocks of suitable dimensions, by joints of right section, formed by intersecting the course by vertical planes through the axis $L'L''$. If $IE_{,}$ and $I_{,}F_{,}$ (*Fig.* 2), for example, be taken as the projections in plan of two joints of right section, the figure $II_{,}F_{,}E_{,}$ will be the projection in plan of a block or voussoir of the course in question. The figure $E'I'I''E''$ (*Fig.* 1) is the projection of the lower conical joint of this voussoir; $F''G''G''F''$ that of the upper conical joint; and $E'F'F''E''$ that of the portion of the soffit. The joints of right section of the adjacent courses break joints, as shown at $E''E'$, $V''v'$, &c., on the curves $V'L''$, $V''L''$, &c. (*Fig.* 1), which are the projections of the circles cut from the soffit by the joints of right section.

Application. Having determined the projections of the bounding lines and surfaces of a voussoir, their true dimensions can be easily determined, and from them the size of a block from which the voussoir can be cut. Taking, for example, the voussoir projected in plan (*Fig.* 2) in $II_{,}F_{,}E_{,}$, from an inspection of the projections (*Figs.* 1, 2) it is obvious that the dimensions of a block from which it can be cut must be such that the figure $II_{,}F_{,}E_{,}$ (*Fig.* 2) can be inscribed within its base, and its thickness be equal to the vertical height between the lines $II'G''$ and $E'E''$ (*Fig.* 1), the total depth of the voussoir. Having selected a block of the suitable dimensions, the different lines and surfaces of the voussoir can be obtained in their true dimensions from its projections (*Figs.* 1, 2, 3), and marked out on the sides and bases of the block.

It will be seen that the end of this voussoir, which forms a portion of the soffit, is comprised between the two meridian planes $IE_{,}$ and $I_{,}F_{,}$ (*Fig.* 2) and the upper and lower conical joints. The points F', F''', E', E'', (*Fig.* 1) are therefore the projections of the four angular points of the voussoir, on the soffit, and lie upon the circumference of a small circle of the dome passing through the points of which these are the projections on *Fig.* 1. This small circle can be readily constructed (*Fig.* 4), since the lengths of the chords joining the two upper points $F'F''$, and the two lower $E''E''$ (*Fig.* 1), are given in their true dimensions $E_{,}F_{,}$ and EF (*Fig.* 2); and the diagonals projected in $EF_{,}$ and $E_{,}F$ (*Fig.* 2) can be readily obtained in their true dimensions. Having set out the small circle determined from these elements (*Fig.* 4), it will limit the portion of the soffit on the end of the voussoir, and will serve as a guide to the workman in working it out.

Fig. 3 gives the true dimensions of the side of the voussoir in the meridian plane $IE_{,}$ (*Fig.* 2).

Prob. 11, *Pl* 6. *To construct the projections and true dimensions of the voussoirs of the gate-recess.*

The line BC (*Fig.* 2) represents the trace of the vertical face or front of a wall; ad that of the back, also vertical. Through this wall an arched gate-way is to be so constructed, that the gate, composed of two leaves, may be placed midway between the face and back, as at AD, and when open the leaves shall be thrown back, taking respectively the positions $A_{,}a$, $D_{,}d$, by revolving around the vertical axes projected in $A_{,}$ and $D_{,}$. In this way, the gate occupies a recess within the wall, from which circumstance the problem is named.

Let B', E', C', (*Fig.* 1) be the curve of right section of the right arch, $BB_{,}C_{,}C$ (*Fig.* 2) its plan. Let the top of the gate when closed be a semicircular cylinder, $A'D'$ (*Fig.* 1) being its diameter, and the rectangle $AA_{,}D_{,}D$ (*Fig.* 2) its plan; the gate when closed shutting against the plane surface ring projected (*Fig.* 1) between the two semicircles; and *Fig.* 2 in the line AD.

The problem to be solved consists in so arranging the surface of the recess under which the leaves swing in being opened or closed: 1st, that it shall offer no obstruction to the play of the leaves; 2d, that it shall be one of easy geometrical construction; 3d, that it shall present a pleasing architectural effect.

The lines $A_{,}a$, $D_{,}d$, (*Fig.* 2) being the traces of the vertical side planes of the recess against which the leaves rest when closed, these planes are each terminated at top by an arc of a circle assumed at pleasure, but of greater radius than $A'L'$ (*Fig.* 1). To construct this arbitrary arc (*Fig.* 3), revolve the side plane $D_{,}d$ around the axis projected in $D_{,}$ (*Fig.* 2), and $D'D''$ (*Fig.* 3), parallel to the face of the wall, into the position $D_{,}d_{,}$. Assume d''' the highest point of the arbitrary arc in the revolved position, at the same height as k' (*Fig.* 1) is above L', and construct an arc passing through $D'd'''$ and tangent to $D'D''$, and let this be taken for the required arc. Supposing the side plane revolved back to its primitive position, $D'd''$ will be the projection of the arc; and $d'd''$ that of the vertical edge of the back and side planes.

Let this arbitrary arc, the semicircle projected in $A'k'D$ (*Fig.* 1) and $A_{,}L_{,}$ (*Fig.* 2), and the axis of the arch projected in L' (*Fig.* 1), Ll (*Fig.* 2), be taken as three directrices of a warped surface, to form a portion of the upper surface of the recess. The projections of the extreme position of the element of this surface will be $d''x'L'$ (*Fig.* 1), day (*Fig.* 2). A like surface covers the opposite side.

To form the top, the two warped surfaces determined are connected by a third, which must be tangent to each of them along the extreme element $d''L'$, $a''L'$ of each, so that the three surfaces may appear as a continuous surface, and thus satisfy the 3d condition. To satisfy this condition, let the axis of the arch and the semicircle, which are two of the directrices of the two first warped surfaces, be taken as two of the directrices of the third. This will give two tangent planes common to the surfaces along each of the elements $d'L'$, $a''L'$. Construct now a tangent plane to the warped surface found at the point d'', by drawing a tangent to the curve projected in $D'd''$ at this point, and through this tangent and the element projected in $d''L'$ passing a plane. The element pierces the vertical plane of which $A_{,}D_{,}$ is the trace at x, x'; the tangent to the curve $D_{,}d''$ at d'' intersects the vertical line $D'D''$ at D''; joining then D'', x', it will be the projection of the trace of the tangent plane on the vertical plane $A_{,}D_{,}$; the projection of its trace on the vertical plane ad is $v'd''w'$, parallel to $x'D''$. Drawing an arc of a circle passing through a'' and d'' and tangent to $v'w'$, if it is taken as the third directrix of the second surface, the two surfaces will be tangent, as they have a third common tangent plane at d''.

The 2d condition is satisfied by taking warped surfaces to form the *soffit*, or top surface.

Having constructed the warped surfaces, with these arbitrary conditions, it will be necessary to ascertain whether they satisfy the 1st condition. To do this, it will be observed that the top of the leaf describes, in its revolution, a surface, and which, to satisfy this condition, should not intersect the warped surfaces within the side plane, as $A_{,}a$, for example. This intersection will be found by the usual methods for finding the intersections of two given surfaces. The line $r's'$ (*Fig.* 1), for example, may be assumed as the vertical trace of a horizontal plane intersecting the two surfaces. This plane will cut from the surface described by the top of the leaf an arc of a circle, projected in sr (*Fig.* 2) and from the warped surfaces an arc $ss_{,}r$; and as these intersect at r, without $A_{,,}a$, the surfaces do not interfere along this horizontal plane. The same construction would be made for other points.

The bounding lines of the top surface being found, the arch is divided off into five equal parts, as $B'E''$, &c. The planes of the voussoir joints pass through the axis of the arch and extend to the points I', G', &c., arbitrarily chosen and from these last points the voussoir joints are vertical.

To determine the true dimensions of the joints $M'N$ and $Q'P'$ (*Fig.* 1); let each of them be revolved around their lower horizontal edges, projected in M', Q', parallel to the horizontal plane; taking $M'N'$, this is done by drawing through M' a line parallel to $L'D'$ and setting off along it, from M', distances equal to $G'F'=M'N'$, $F'i$, $F'h'$, and, to simplify the construction, drawing through the points thus set off, lines parallel to $L'M'$ to intersect $L'D'$; from these last points drawing lines parallel to $L'l$ the points $Lk'E'i'h'G'I$ (*Fig.* 4) will be found, which joined will give the figure and true dimensions of the joint through $M'N'$. In a like manner the figure $Lk'E'i'f'g'G'I$ and true dimensions of the joint through $Q'P'$ are obtained.

Application. To cut the voussoir out of a block of the form of a rectangular parallelopiped, its dimensions must be such that the figure $N'M'Q'P'O'$ (*Fig.* 1) can be inscribed within the end, and its length be equal to IG' (*Fig.* 4). Having set out the bounding lines of the different surfaces from *Fig.* 1, the plane and cylindrical portions will be first cut off; next the portion of the warped surfaces, by first working down to the positions of several of the intermediate elements, determined from the drawing, and then finishing off by the eye the portions of the surface between these elements.

Prob. 12, *Pl.* 7. *To construct the projections and true dimensions of the steps of the geometrical stairway.*

Let $ABCD$ (*Fig.* 1) be the polygonal base of a vertical wall, along which a flight of stone steps is to be built. Let XYZ, xyz, be two curves having the relation of involute and evolute to each other; the one XYZ being the base of a vertical cylinder, the surface of which limits the ends of the steps, and which is termed the *well* of the stairs. Let $X_{,}Y_{,}Z_{,}$ be another curve parallel to the one XYZ, and at the distance from it that persons going up or down the stairs would naturally take; and where, on this account, the top of each step, or the *tread* should have a uniform breadth. This tread added to the height, or *rise*, being usually assumed at twenty-two inches, as a convenient distance for each step.

The problem consists: 1st, in arranging the tread and rise with these conditions; 2d, in making the under surface of the stairway a helicoidal surface; 3d, in arranging the joints between the steps so as to be plane surfaces, and normal to the helicoidal surface at the middle point; 4th, in determining from these conditions the form and dimensions of each step.

Having set off the equal arcs X 1, 1-2, 2-3, &c. (*Fig.* 1) along the curve $X_, Y_, Z_,$ equal to the assumed tread, draw, through these points, lines tangent to the curve *xyz*, and prolong them to the line *bcd* parallel to BCD; the quadrilaterals thus formed, between XYZ and *bcd*, will be the true dimensions of the top surface of each step. Midway between the equal arc, as at $Y_,$, $Z_,$, &c., draw lines also tangent to *xyz*, and let these be assumed as the projections of the edges of the joints along the helicoidal surface; and, to fix their position, let the edge of each joint be taken at the distance of half the rise below the top of the step. The points thus determined will lie on a helix, which at each of these points is half a rise below the top of the step, and the inclination of the tangent to which at any point will be the rise or height of each step divided by the uniform tread.

Having fixed the position of the helix, the helicoidal surface is generated by moving a right line along it so as to be parallel to the horizontal plane, and, in all of its positions, be projected normal to the curve XYZ.

Let $y_,$, the middle point of the lower edge Ym, be taken as the point at which a normal plane is to be passed to the helicoidal surface for the joint in question. This plane is determined as shown on *Fig.* 2, by *Prob.* 7 (*Pl. A, Fig.* 6), and in like manner the one at $z_,$ on *Zo* as shown in *Fig.* 3.

Having determined these planes, their intersections with the tops of the steps will give the lines Nn, Op, parallel to Ym, Zo, which are the top edges of the joints.

With the *data* now determined, the form and dimensions of the ends of the step to which these two joints belong can be determined. The larger end of the step is contained in a vertical plane of which *mq* is the trace on the horizontal plane. Draw a line $B'C'$ parallel to *bc* and at any assumed distance from it; this may be taken as the revolved position of the top line of the end, on the horizontal plane. From the points m, n, o, p, and q draw perpendiculars to *mq;* set off on these the distance $q''q'$ for the rise of the step; $m''m'$ equal to half a rise; o' half a rise below p'; join the points thus set off. The figure $m'n'q''q'p'o'$ is the one required. To find that of the other end (*Fig.* 5), the portion of the cylinder of the well YM is developed, and the corresponding points set off on it; the figure $Y'N'M''M'O'Z'$ is the one required.

Application. To cut the stone, a block must be taken upon the top of which the figure *YmqM* (*Fig.* 1) can be set off, and on the large end *Fig.* 4. Having dressed off the plane and cylindrical surfaces, the portion of the warped surface can be dressed off as in the preceding problem.

Prob. 12, *Pl.* 8. *To construct the projections and true dimensions of the voussoirs of the groined annular and radiant arch.*

The soffit of this arch is formed of the surfaces of an annular and radiant arch, the intersections of which form the curves of the groin.

To find the horizontal projections of these curves, let (Fig. 2, Pl. 8) B′ C′ be the diameter, and K″ the centre of a semicircle, contained in a vertical plane passed through the vertical line projected at L horizontally. If the semicircle be revolved around the vertical L it will generate the surface of an annular arch.

From the point L (Fig. 1) in the horizontal plane of the diameter B′ C′ of the semicircle let two lines L B′ and L c be drawn, making any convenient angle with each other, and let the chord of the arc B′ c, included between them, described by the point B′ of the semicircle, in its revolution, be taken as the transverse axis of a semi-ellipse, contained in a vertical plane of which B′ c is the trace; the conjugate semi-axis of this ellipse being equal to the radius of the semicircle. Then if a right line be so moved that it shall, in all of its positions, be parallel to the horizontal plane of the semi-transverse axis of the ellipse, intersect the vertical through L, and rest on the curve of the semi-ellipse, it will generate the soffit of the radiant arch.

The horizontal projections B′ L, C, and C′ L c, which are those of the groins, will be obtained, by finding the intersections of the projections of the corresponding elements cut from the two soffits by horizontal planes. The point L₁ being that of the highest point; and the points B′ C′ C, c being those of the lowest points of the groins.

If now the semicircle be divided into five equal parts, and the right sections of the voussoirs of a cylindrical arch be drawn (Fig. 2); the joints E′ I′, F′ G′, &c., of this right section, in the revolution of the vertical plane containing them, will describe zones of conical surfaces, the vertices of which will be on the vertical through L, where these joints prolonged intersect it. In like manner, the lines I′ II′, G′ G″, &c., will describe cylindrical surfaces, having the same vertical for their common axis; and the horizontal lines, as G′ H′,

&c., will describe zones of circles. Thus completing the bounding surfaces of the voussoirs belonging to the annular arch.

As the soffit of the radiant arch is a right conoid, having the horizontal plane, containing the semitransverse axis of the directing semi-ellipse, for its plane director, and the vertical through L as its right line directrix, its joints, along the rectilinear elements corresponding to the horizontal circles described by the points Q', M', F', E', of the semicircle, to be normal throughout to the soffit, would require to be warped surfaces. To avoid the inconvenience of constructing these, a plane surface joint is substituted for each stone instead of the other; and this is so taken, that it shall be normal to the soffit at the middle point of that portion of the right line element of the soffit which belongs to the joint. Taking for example the element projected in $L E_1 e_2$, and assuming that the groin voussoir is bounded, on the radiant portion of the arch, by the lines $F_1 f_1$; by f_1 I, which is the projection of the line cut from the soffit by a vertical cylinder; by the projection $F_1 E_1$ of the groin curve, and by the line $E' e_2$; then $F_1 f$ will be the lower edge of the radiant joint, corresponding to the lower edge of the annular joint, described by the point F'; and $E_1 e_2$ will be the lower edge of the joint below, corresponding to the one $E' I'$.

Having the lower edge $E_1 e_2$ of the joint, the other bounding lines of it are found by constructing a normal plane to the soffit containing the right line element through E_1, at the middle point e of $E_1 e_2$, and finding the intersections e, I of this plane with the vertical cylinder f_1 I that limits the voussoir; with the conical zone described by $E' I'$, and which is projected in $E_1 z$; and finally with the horizontal plane which passes through I' (Fig. 2), and which will be projected through I parallel to $E_1 e_2$.

In like manner the plane joint projected in $F_1 f_1 g_1$ G can be found.

The portion of the groin stone, belonging to the annular arch, is limited by a vertical plane passed through the points L, F_2, H; the line $F_1 E_1$ of the groin; the upper and lower conical joints; the cylindrical surface projected in z H; and the horizontal plane through $G' H'$ (Fig. 2).

Note.—To construct the normal planes of the plane joints of the radiant arch see Prob. 6.

All the horizontal lines of the surfaces bounding the groin voussoir in question, are projected in their true dimensions in Fig. 1.

Having found the horizontal projections of all the lines

bounding the groin stone considered, the true dimensions of all the developable surfaces by which it is bounded
can be found by methods already used in the preceding
Probs.

Take, for example, the plane joint of the radiant arch,
projected in E_1 e_2 I z. Having first determined the tangent
plane to the middle point e_2 of the lower edge of the joint,
and its trace t_2 t_2 on the horizontal plane of the springing
lines, by Prob. 6, let this last plane be intersected by a vertical plane $X Y$ perpendicular to the lower edge E e_2 (Fig. 1),
and let it then be transferred, parallel to itself, to $X Y$ (Fig.
3). The plane $X Y$ will cut from the tangent plane a line
which, in the revolved position of the plane $X Y$ (Fig. 3),
will be projected in t_2 E'; E'' E' (Fig. 3) being equal to E''
E' (Fig. 2). From E' (Fig. 3) drawing a line perpendicular
to E' t_2 it will be a normal to the soffit of the radiant arch at
e_2 (Fig. 1), and where this normal intersects at I' a line parallel to E'' t_2, and at the same height above it as I' (Fig. 2)
is above E'', the line E' I' (Fig. 3) will be the true width of
the plane joint considered. Now, revolving this plane joint
around the line E' I' to coincide with the vertical plane $X Y$
(Fig. 1), the points e_2, E_1 (Fig. 1), for example, will fall in a
perpendicular to E' I' (Fig. 3) as far from it, at e'' and E'',
as they are in horizontal projection from $X Y$ (Fig. 1). In
like manner, the points i'' and I''' (Fig. 3) are found; and
e'' E'' I''' i'' will be the true dimensions of the plane joint.
E'' I''' will be the intersection of the plane radiant joint with
the corresponding conical joint of the annular arch; and e'
i' the intersection of the same joint with the cylindrical joint
f_1 I of the radiant arch.

Fig. 4, showing the true dimensions of the upper plane
joint, is constructed by a like series of operations.

Fig. 5 is the development of the cylindrical joint of the
radiant arch projected in f_1 I; and of the cylindrical surface
of the groin stone of the annular arch projected in z II.

The projections of the conical joints of the annular arch
are easily found, by developing the cones to which they belong.

Fig. 6 is the development of the end surface of the exterior
voussoir of the radiant arch which joins the groin voussoir
considered.

These arches rest, as in the cylindrical groined arch, on
pillars. The tops of these pillars, on a level with the springing lines of the arches, are shown in the trapezoids B B' a' a,
C C' b' b, &c. (Fig. 1.)

The dimensions of the block for the groin stone in question are all given in the projections, sections, &c., of Figs. 1, 2, 3, 4, 5, and the developments of the conical joints. With these elements, the bounding lines can be marked out on the block, and the voussoir be worked off, by methods similar to those pointed out in the two preceding problems.